VIOLENT
POLITICS

ALSO BY WILLIAM R. POLK

The Birth of America

Understanding Iraq

The United States and the Arab World

The Elusive Peace: The Middle East in the Twentieth Century

Neighbors and Strangers: The Fundamentals of Foreign Affairs

Polk's Folly: An American Family History

Out of Iraq: A Practical Plan for Withdrawal Now
(with Senator George McGovern)

VIOLENT POLITICS

A HISTORY OF INSURGENCY, TERRORISM & GUERRILLA WAR, FROM THE AMERICAN REVOLUTION TO IRAQ

WILLIAM R. POLK

HARPER

An Imprint of HarperCollins*Publishers*
www.harpercollins.com

HarperCollins books may be purchased for educational, business, or sales promotional use. For information, please write: Special Markets Department, HarperCollins Publishers, 10 East 53rd Street, New York, NY 10022.

FIRST EDITION

Designed by William Ruoto

Library of Congress Cataloging-in-Publication Data is available upon request.

ISBN: 978-0-06-123619-8
ISBN-10: 0-06-123619-5

07 08 09 10 11   OV/RRD   10 9 8 7 6 5 4 3 2 1

For Elisabeth with love

ACKNOWLEDGMENTS

THIS BOOK IS THE PRODUCT OF OBSERVATION, RESEARCH, AND DIS-cussions that go back to my visits to the then Palestine Mandate in 1946. Over the years, I have had occasion to visit the scenes of guerrilla wars there and in Greece, Cyprus, Lebanon, Iraq, Vietnam, Sri Lanka, Kenya, Yemen, the Sudan, Nigeria, Morocco, Algeria, Ireland, and the Basque country. During those visits I was helped toward understanding by hundreds of participants, officials, peace seekers, and observers.

I also greatly profited from my time as a member of the Policy Planning Council of the U.S. Department of State, where my colleagues constituted a permanent seminar on world affairs. The chairman, Walt Rostow, was focused on the Vietnam War and was a generous and fair-minded counselor and friend. It was because of our disagreements that I did my first formal study of guerrilla warfare, which I presented at the National War College in 1963. With Rostow's support, the urging of McGeorge Bundy, then head of the National Security Council staff, and at the request of Governor Chester Bowles, then under secretary of state, I was able to spend some time in a number of actual or potential "trouble spots," including Afghanistan, Yemen, Pakistan/India, and the Horn of Africa. As a member of the council, I served on various committees (task forces as we then called them) on a wide variety of foreign policy issues and chaired the group devoted to the Franco-Algerian war. My position gave me access to the wide variety of materials collected by government agencies and analyzed by such groups as the department's Bureau of Intelligence and Research and the Central Intelligence Agency's Office of National Estimates. The heads of those two organizations, Thomas Hughes and Sherman Kent, were among the most able and intelligent scholars I have ever known. David Halberstam rightly emphasized that, as a group, these people were America's Best and Brightest.

When I left government to become professor of history at the University of Chicago and subsequently president of the Adlai Stevenson Institute of International Affairs, I was able to build a sort of pocket university on the problems of violent politics, drawing on a remarkable array of skilled men and women including Murray Kempton, David Halberstam, Neil Sheehan, Eqbal Ahmad, General Indar Rikhye, Kenneth Hansen, Lord Caradon, Dr. William Grier, Dr. Anthony Storr, Manfred Halpern, General Fred Haynes, Fred Arkhurst, U Thant, and Richard Barnet.

Over the years, my children, Milbry, Alison, George, and Eliza; my wife, Elisabeth, and many friends have been warm supporters and helpful critics. Many former colleagues, friends, and acquaintances, especially Ambassador Robert Keeley and John Whitbeck, have also acted as an informal intelligence service, assisting me to access the world press; without their help, my study would have been harder and my account poorer. Dr. Hans Noll illuminated one of the baffling calamities of the Iraq wars, its unclear dimension.

Lastly, I have been generously supported in my research by the Rockefeller Foundation, especially by the late John Marshall, the Guggenheim Foundation, the Ford Foundation, and the W. P. Carey Foundation, to whose founder, William Polk Carey, I am especially grateful.

. . . to observe guerrilla warfare is to find yourself everywhere
and nowhere, witness to a particularly fragmented reality.
It helps a great deal if you can rely, in your evaluation of a
resistance movement, not only on what you observe in the field
and on whatever information from other witnesses that you are
able to verify but also on an accumulated knowledge of other
guerrilla struggles.

> —Gérard Chaliand, *Report from Afghanistan* (*Rapport sur la
> résistance afghane*), 1981

[American Indians] never congregate in a group that could be
destroyed by the superior firepower of the regular troops and
when attacked they never stand their ground, but immediately
give way only to return to the charge when the attack ceases.

> —Colonel Henry Bouquet, *Indian Warfare,* 1763

Let us approach this pivotal fact in a humble yet hopeful
mood—
    We have had no end of a lesson, it will do us no end of
good!

> —Rudyard Kipling on the Boer War, 1902

Wars of opinion . . . result either from doctrines which one
party desires to propagate among its neighbors, or from
dogmas which it desires to crush—in both cases leading to
intervention. Although originating in religious or political
dogmas, these wars are most deplorable; for, like national wars,
they enlist the worst passions, and become vindictive, cruel and
terrible.

> —General Antoine Henri Jomini, *The Art of War,* 1838

At first, there is a partially armed band that takes refuge in
some remote, hard-to-reach spot. It strikes a lucky blow against
the authorities and is joined by a few more discontented
farmers, young idealists, etc. It reconnoiters inhabited areas,

contacts residents, and conducts light hit-and-run attacks. As new recruits swell the band, it takes on an enemy column and destroys its leading elements . . . Next, the band sets up semipermanent encampments, establishes service echelons, and adopts the characteristics of a government in miniature. Small industries, hospitals, and radio stations are set up, laws are decreed, a court administers justice, and ideological indoctrination is intensified. An enemy attack is beaten off, more arms captured, and more guerrillas armed . . . [eventually] guerrilla combat forces capture heavier arms and begin positional warfare.

    —Che Guevara on guerrilla warfare, 1960

Without a political goal, guerrilla warfare must fail, as it must if its political objectives do not coincide with the aspirations of the people and their sympathy, cooperation, and assistance cannot be gained.

    —Mao Tse-tung, *Yu Chi Chan,* 1937

When we speak of the guerrilla fighter, we are speaking of the *political partisan,* an armed civilian whose principal weapon is not his rifle or machete, but his relationship to the community, the nation, in and for which he fights . . . It is his camouflage, his quartermaster, his recruiting office, his communications network, and his efficient, all-seeing intelligence service.

    —Robert Taber, *War of the Flea,* 1965

[Guerrilla warfare requires] serious study of all historical experience. Although a wealth of material existed then [in the 1950s] and much more has since been developed, no such study has yet been undertaken in this country . . . In Indochina and Cuba, Ho Chi Minh and Ernesto (Che) Guevara were more assiduous.

    —Brigadier General Samuel B. Griffith, USMC, 1961

CONTENTS

Introduction xiii

1 The American Insurgency 1
2 The Spanish *Guerrilla* Against the French 20
3 The Philippine "Insurrection" 35
4 The Irish Struggle for Independence 54
5 Tito and the Yugoslav Partisans 72
6 The Greek Resistance 89
7 Kenya and the Mau Mau 107
8 The Algerian War of National Independence 124
9 The Vietnamese Struggle Against the French 144
10 America Takes Over from France in Vietnam 163
11 The Afghan Resistance to the British 181
   and the Russians

Conclusion: The Very Expensive School 202
Notes 225
Index 261

INTRODUCTION

FOR MORE THAN A CENTURY, INSURGENCY HAS BEEN A FOCAL POINT in foreign affairs; its principal tactics—terrorism and guerrilla warfare—have been employed from Indonesia to Ireland and from Colombia to China. People of all races, cultures, climates, economies, religions, and languages have engaged in it. Since it is so widespread, and so costly and painful for all its participants, we must ask why it is so prevalent.

We should begin by noting what is common to all insurgencies. No matter how they differ in form, duration, and intensity, a single thread runs through them all: opposition to foreigners. This propensity to protect the home community arises because all human beings are territorial. Among primitive peoples territoriality is immediate and physical. It is usually expressed by a kin group or clan living together in a defined area small enough that the members are in daily contact. Anthropologists have found that such groups seldom exceed a hundred individuals. They are people who know one another intimately, live side by side, hunt or herd together, and often own their property in common. They are bonded by the absolute imperative to protect the resources that sustain their lives. Today such groups are rare, but the way they live is how all our ancestors lived for millions of years, conditioned over millennia to turn inward to rally support and outward to repel intruders.

As people settled down, first to become farmers, then townsmen, and finally to amalgamate into nations and even larger agglomerations, as most living people have done, territoriality has become more abstract, so that people who are not bound together by kinship may identify with one another without close physical contact as fellow members of an emotional, ideological, religious, or cultural "neighborhood."

Thus, at one extreme, the operative society is a single clan; at the other, millions of people may form a nation or ethnic or religious group that takes on some of the sense of cohesion so evident in a small kin group. Regardless of size or definition, however, the society is the touchstone of individual identity. Consequently, while members may accept outsiders as guests or "guest workers" and even on occasion allow them to join, they rarely and usually only temporarily are willing to accept foreigners as rulers. The very presence of foreigners, indeed, stimulates the sense first of apartness and ultimately of group cohesion. This is a process that is partly automatic but is often fostered both by native leaders and by the actions of their foreign enemies. Natives tend to react violently if they perceive that foreigners are changing or corrupting their way of life, while the imperial or colonial government seeks to suppress their rebelliousness. In the action and reaction, what I have called the climate of insurgency is created. As each loses legitimacy in the eyes of the other, both employ violence. The government uses force to attempt to overawe native opponents, to create stability, and to prevent lawlessness. For the natives, who cannot otherwise convince the foreigners to leave, insurgency becomes the politics of last resort.

Is insurgency really that consistent? No, obviously there is considerable variation. In the following chapters, I will show how different cultures, ideologies, and degrees of political consciousness among insurgents—and also among their foreigner overlords—have shaped the nature of struggles all over the world during the last three centuries. But from this analytical history will emerge, I argue, that they do so without altering the fundamental motivation of violent politics. That motivation and the course of action to which it gives rise aim primarily to protect the integrity of the native group from foreigners. That the heart of insurgency is essentially anti-foreign is the central thesis of this book.

I believe authors owe their readers an account of how they arrived at their analysis. Here is mine.

From the summer of 1961, when I became a member of the Policy Planning Council of the U.S. Department of State, I was given access to the full range of information flowing into the government. My first assignment that would eventually influence this book was to head the interdepartmental task force monitoring and preparing to deal with the Algerian guerrilla war against the French. I will describe what I then learned in Chapter Eight. Algeria was an eye-opener, but even before entering government service, I had already observed, sometimes quite closely, several guerrilla wars. So it was natural that I paid particular attention to the growing conflict in Vietnam.

In Algeria, America played a minor and peripheral role, but in Vietnam, the State Department, Agency for International Development, Central Intelligence Agency, Department of Defense, and the separate armed services, as well as journalists and nongovernmental organizations, began to accumulate enormous collections of facts. No country was ever so reported upon as was Vietnam by Americans. From 1962, shortly before the major escalation of American activities began, I spent part of each day perusing the deluge of cables, intelligence reports, and summaries that poured across my desk on Vietnam. Of these a small fraction subsequently ended up in *The Pentagon Papers*.

In all the mass of materials, thousands upon thousands of pages, however, one looked in vain for a satisfying definition, much less a coherent and penetrating analysis, of guerrilla warfare. Everything I came across was episodic, without historical depth, short on questions but quick with answers that often seemed unrelated to events. As the months passed, I came to believe that our lack of any evident means to make sense of the daily events was so dangerous—already we were fairly deep into Vietnam and obviously were headed deeper—that

I took six weeks off from my regular tasks at the Policy Planning Council and read everything I could find on insurgency.

Learning about my study, the National War College invited me to summarize my findings for its graduating class of the "best and brightest"—army, air force, and marine colonels and navy captains—who were headed for senior command. I was delighted, as the occasion would force me to try out what I had learned on a very critical audience, one preparing for combat in Vietnam. Before them, I then argued that guerrilla warfare was composed of three elements—politics, administration, and combat. I had concluded that combat was the least important element in the struggle. I told the audience that we had already lost the political issue—Ho Chi Minh had become the embodiment of Vietnamese nationalism. That, I suggested, was about 80 percent of the total struggle. Moreover, the Viet Minh or Viet Cong, as we had come to call them, had also so disrupted the administration of South Vietnam, killing large numbers of its officials, that it had ceased to be able to perform even basic functions. That, I guessed, amounted to an additional 15 percent of the struggle. So, with only 5 percent at stake, we were holding the short end of the lever. And because of the appalling corruption of South Vietnamese government, as I had a chance to observe firsthand, even that lever was in danger of breaking. I warned the officers that the war was already lost.

The War College audience was disciplined to be polite, but the officers were furious. They knew how to fight and were anxious to prove their skill and determination. Vietnam was the prime opportunity of their professional careers. The idea that we would lose Vietnam was regarded in 1963 as rank heresy. Moreover, I was just a civilian. They were soldiers. I was dealing only with theory; they would make facts on the ground. That we would lose the war in Vietnam was unthinkable. The officers were not alone in this conclusion. The chairman of the Policy Planning Council, my boss Walt Rostow, who later as head of the National Security Council came to be regarded as the architect of our Vietnam policy, admonished me for being a Cassandra. I accepted the designation; after all, Cassandra had been right. Although we remained friends, we never resolved our differences over Vietnam, and in due course I resigned.

But then a curious thing happened. In 1967, when I had become president of the Adlai Stevenson Institute of International Affairs and professor of history at the University of Chicago, I was invited to lecture again at the War College on more or less the same topic. Expecting a hostile audience of the current class of senior officers, I hedged my remarks somewhat by saying that my intent was to provoke thought. Thereupon an army brigadier general stood up and said that he did not understand my diffidence. Laughing, I recounted my previous experience, to which he replied, "But we have all been there now."

As president of the Adlai Stevenson Institute of International Affairs, while I did not myself speak often about Vietnam, I arranged fellowships that enabled David Halberstam to write his book on the formation of American policy, *The Best and the Brightest*; Neil Sheehan to begin his study of counterinsurgency that led to his book *A Bright and Shining Lie*; and Richard Pfeffer to bring together in a conference the leading critics and proponents of American activities in Vietnam, from whose writings and papers he edited *No More Vietnams? The War and the Future of American Foreign Policy*.

Like many observers, I hoped that Vietnam would be the final lesson for Americans that no matter how many soldiers and civilians were killed, how much money was spent, how powerful and sophisticated were the arms employed, foreigners cannot militarily defeat a determined insurgency except by virtual genocide. We came close to genocide in Vietnam—where we dropped more bombs than all the armed forces of the world exploded during the Second World War, poisoned or burned vast tracts of the country, and killed about two million people. Despite all this, we still lost the war. We did not learn the lesson in Vietnam. We still have not.

According to the *Oxford English Dictionary*, "guerrilla," the Spanish word for "small war," was introduced into English in an 1809 dispatch from the Duke of Wellington during the Anglo-Spanish campaign against Napoleon's army. With unintended irony, the editors of the *Dictionary* commented that "it is now somewhat rare."

Now somewhat rare! Forty years later with perhaps $2 trillion wasted and about thirty thousand Americans and millions of people of other lands wounded or killed, books on Spain, Cuba, Colombia, Ireland, the Basque country, Cyprus, Afghanistan, Çeçnya, the Uighur areas of China, Sri Lanka, Burma, the Philippines, Malaysia, East Timor, the Sudan, Somalia, Yemen, Vietnam, Algeria, Iraq—I cannot list them all—fill my shelves. It is obvious just from the ones I have named that insurgency and its most dramatic manifestations, guerrilla warfare and terrorism, are worldwide phenomena of the greatest importance, yet very few of the authors of books on current conflicts offer insightful social, political, cultural, or economic analysis.

Those who set out to analyze the nature of insurgency emphasize—many exclusively—the military aspects. These, as I mentioned and will show in the following chapters, are the least important. Many concern themselves only with weapons; some even just with uniforms. They seem to believe that the symbols shape the movements. Then there are the eyewitness accounts that attempt to give a vision of battle or the life of combatants. Rarely are there meaningful accounts of the history of the people or any attempt to understand their societies and cultures. Yet these, after all, are what give rise to the conflicts.

I have cast my account in eleven chapters, each designed to show how a particular climate of insurgency was created, how an outbreak of violence was triggered, the stages through which the insurgents progressed, and finally the outcome of each struggle. I have adopted this approach because I want to show both the ways in which various peoples—Muslims, Christians, Jews, Hindus, Buddhists, and atheists; Africans, Asians, Americans, and Europeans; advanced societies and the less developed—have adopted terrorism and guerrilla warfare when they decided that other means of politics were unavailable.

I will also use the historical record to bring out what I see as the stages through which insurgencies evolve. Each of the insurgencies with which I deal—and others that are too numerous to be fully recounted here—begins with almost ludicrously tiny groups of disaffected people who sally forth against vastly superior armies and police forces. The odds appear impossible—even absurd. Consider the record:

In this preliminary stage, the insurgents are too few to fight as guerrillas so they fight as terrorists. This was what happened in Cyprus, where fewer than eighty men attacked the ruling British colonial government that was supported by thousands of troops and police. In the Palestine Mandate, a tiny group to be known as Irgun Zeva'i Le'umi split off from the Jewish Agency's military force, the Haganah, to attack the British administration; briefly during the Second World War, the Irgun ceased attacks on the British and Avraham Stern split off to form a new group, known as Stern or LEHI, and gathered a handful of followers, probably fewer than twenty, to continue attacks on the British. Stern's most spectacular action was the murder of the senior British official in the Middle East in 1944. Stern was so successful that Irgun, by then under the leadership of Menachem Begin, swung back toward Stern's policy and in 1946 blew up the largest hotel in Jerusalem, where a number of senior British officials were housed. By that time, Irgun and Stern numbered perhaps a hundred men. Most other groups were similarly tiny at inception. In Yugoslavia, for example, Drazha Mihailovic (also written Mihailovich), a Serb who had been a colonel in the defeated royalist army, began the Cetniks with only twenty-six like-minded former officers and soldiers. In Greece, the resistance was formed by only fifteen men. Then, in Vietnam, the first action by the Viet Minh against the French in 1944 involved their total force—only thirty-four Vietnamese. Such small groups could not engage in guerrilla warfare. For them, acts of terrorism were the only possible acts. So terrorism is often the first stage of insurgency.

As terrorist acts succeed, other angry men and women join or form similar small groups. When the dominant government seeks to suppress them, two things frequently happen. Almost inevitably the government disrupts the lives of innocent bystanders and hurts or kills still more. In 1808 in Spain Napoleon's soldiers routinely hanged all rebels they caught and those suspected of favoring them. The relatives and friends of the hanged quickly came to hate the French. Against the Philippine rebels, first the Spaniards and then the Americans undertook search and destroy operations that killed thousands of people in the 1890s, tortured or humiliated many more, and burned scores of villages. Doing so triggered Philippine resistance. In Yugoslavia during the Second World War, the Germans employed a draconian system of reprisals, executing not only all the partisans they captured but hundreds of civilian hostages in retaliation for the death of each German soldier. The relatives, neighbors, and friends of those killed by foreign troops sought revenge, and the place to get it was in the ranks of insurgents. So from a handful, their numbers grew.

On Cyprus, the insurgent force tripled in a year, from about 80 to 273 active combatants backed up by perhaps three times that many part-time fighters. Castro overthrew the foreign-supported Batista dictatorship with a rebel force that grew from a dozen or so to a force of about 1,500 within a year. The Viet Minh military force grew from 34 in 1944 to about 5,000 in just a few months. In Algeria, the guerrillas began with fewer than a hundred and ultimately reached about 13,000 in their fight against 485,000 French and Foreign Legion soldiers.

Numbers were important because they made possible the spread of insurgency and also because they attenuated the forces of their opponents, whom they forced to protect more territory. More important than numbers, however, was that in their operations, the insurgents came to symbolize the nationalist cause. Mao thought that identifying with the people was the single most important task of the guerrilla. Tito captured the position of national leader because, unlike Mihailovic who temporized, husbanded his resources, and even made deals with the Italians and Germans, he fought. In Vietnam, similarly, so firmly did Ho Chi Minh come to symbolize the nationalist cause that the Americans realized that he would win perhaps 80 percent of the vote in a free election even in the American-controlled south.

Thus, we can observe that after shadowy beginnings, the insurgency in what we can call Phase 1 both gains a critical mass for extended operations and achieves recognition as the national champion. This, as the record will demonstrate, is the most critical event in insurgency. In Vietnam, that event occurred long before the first American soldiers arrived.

However, some insurgencies never progress beyond this first, terrorist, phase. The reasons are several. Among them, perhaps the two most likely are that the original group fails to capture the aura of legitimacy or leadership and so cannot recruit enough followers, or that it cannot find sufficient space for maneuver into true guerrilla warfare. Both of these factors affected the Irish Republican Army (IRA) and the Basque Separatist Movement (ETA).

Toward the end of the first phase, groups of men, often numbering only a score or so, begin to seek targets that will yield what they most desperately lack: weapons. Often, at this time, the insurgents have only shotguns or even just agricultural tools with which to fight; so what the seizure of a small outpost can yield, rifles or even a machine gun or two, is literally priceless. The attackers pick isolated targets where they can achieve superiority. The booty they seize enables them to attract more followers, while the arms they acquire give them more capability. If they are beaten off, as certainly many have been, they run, drop their weapons, and hide. If they have prepared the neighborhood, they can pretend to be just peasants working in the fields. To use Mao Tse-tung's famous phrase, they behave as fish and seek shelter and sustenance among the people, the sea.

Phase 2 of an insurgency comes about when the combatants disrupt the administration of the dominant power and its local allies. The French journalist Bernard Fall searched the French police records to discover the way this happened in South Vietnam. There the local branch of the Viet Minh, the Giai Phong Quan, systematically murdered government-appointed village officials. Fall estimated that the Giai Phong Quan killed about seven hundred during 1957–1958, twenty-five hundred from 1959 to 1960, and four thousand from 1960 to 1961. But it was not just the officials who were liquidated. As George Carver of the CIA wrote in 1966 in *Foreign Affairs*, "The terror was directed not only against officials but against all whose operations were essential to the functioning of organized political society, school teachers, health workers, agricultural officials, etc." Thus, by the time the Americans became active in Vietnam, the South Vietnamese government had virtually ceased to function. It could not collect taxes or even deliver mail much beyond downtown Saigon. Its officials could move only during daylight. Even in Saigon, as I witnessed one night, government patrols avoided the streets when darkness fell.

Disruption is followed by substitution. Having killed or chased away the representatives of the regime, the insurgents immediately begin to create an alternative administration or "anti-state." We can see this most recently in the struggle of the Viet Minh against the French, but it was evident in the American Revolution, where "Committees of Safety" became local governments; the Spanish war against the French, where *juntas* replaced both the old royalist government and the French invaders. Tito, inspired by having observed peoples' committees, known as soviets, during the Russian Revolution, created *odbors* (people's councils) to manage local affairs; the committees opened schools, published a newspaper, and even arranged sports events. Similarly in Greece the National Liberation Front (Ethnikon Apeleftherotikon Metopon, or EAM) underground created village councils that collected taxes to support the *andartes* (combatants). Each committee was headed by a "responsible person" (Greek: *Ipefthinos*), a local man who was almost invariably a member of the Communist Party. In turn, each *Ipefthinos* became a member of the council of a group of villages and so on to create a political pyramid that effectively ruled occupied Greece.

This process is the guerrilla version of what the leading French counterinsurgent, Colonel (later General) Joseph Gallieni, called the *tache d'huile*, or oil spot. Gallieni argued that as each "spot," each village or district, was secured, it would merge with other spots until the whole country was under control. The theory was sound, but the French found that it was easier for guerrilla oil spots to spread than for those of counterguerrillas. Americans, in their turn in Vietnam, found the same.

Why is this? Mao Tse-tung and other guerrilla leaders provide an answer. It is that control of territory has exactly the opposite meaning for guerrillas and counterinsurgents. When counterinsurgents acquire territory, they create new targets for the guerrillas, and so to protect their installations, they have to spread out their forces and go on the defensive. Thus, in any given place, the guerrillas are apt to be able to concentrate to attack with overwhelming force. When guerrillas move into an area, they are not trying to protect installations; their objective is to take control and win over the people. To prevent them from doing so, the counterinsurgents are forced into repressive and unpopular actions. In Malaya first and then in Vietnam, experiments were tried to so isolate the people that the rebels could not reach them. In Malaya the British even prevented villagers from cooking their own food so that they could not pass any of it to the rebels; in Kenya the British put about 150,000 people in concentration camps; in Vietnam the Americans forced virtually the entire rural population to move out of their villages into strategic hamlets, *ap-chien-luoc*. As the editors of the *Pentagon Papers* wrote, "The long history of these efforts were marked by consistency in results as well as in techniques: all failed dismally."

Occasionally, however, neither the guerrillas nor their enemies managed to create organizations outside of small autonomous areas or even single villages. This was the situation in Afghanistan during the resistance to the Soviet Union. The lack of unity made it impossible for the Russians to defeat the guerrillas, but also made it impossible for the guerrillas to defeat the Russians. It left as its legacy the warlords who destroyed much of their own country after the Russians left and created the conditions that made the Afghans welcome the Taliban. It is a legacy that still shapes Afghan society and politics.

Sometimes, in the rebel-controlled areas, even those that were only temporarily secured, quite sophisticated organizations were created. Tito, whose Partisans were nearly always on the run from the far superior German forces, used one period of relative security to organize factories to manufacture rifles and even to turn out cigarettes. Later in the war, his Partisans ran a postal service on a captured railroad. The Viet Minh created repair shops for the weapons they captured and even manufactured duplicates of captured light machine guns. In wartime Greece the EAM organized a system of taxation and countrywide reconstruction support. In Spain, during the war against Napoleon, one guerrilla organization took over and ran the customs for external trade. Although far less sophisticated, the Mau Mau guerrillas in Kenya set up shops in the jungle to convert door bolts and springs into the crude firearms they used against the British. In short, during Phase 2, those insurgent organizations that ultimately were successful created anti-administrations, anti-economies, and ultimately anti-governments for their increasingly large groups of fighters and even larger groups of supporters.

Up to this point, the tactics that have served the guerrillas—hit and run, ambush and fade away, wear down the enemy piecemeal, and concentrate on winning over the general population—no longer seem sufficient. In a memorable passage the American scholar Robert Taber compared the guerrilla to a flea: "The guerrilla fights the war of the flea. The flea bites, hops, and bites again, nimbly avoiding the foot that would crush him. He does not seek to kill his enemy at a blow, but to bleed him and feed on him, to plague and bedevil him, to keep him from resting and to destroy his nerve and his morale. All of this requires time. Still more time is required to breed more fleas. What starts as a local infestation must become an epidemic, as one by one the areas of resistance link up, like spreading ink spots on a blotter."

The insurgent commanders lose patience. They worry that their supporters will tire or even go over to the enemy. Their men age or get wounded or killed. By this time, guerrilla formations have reached a comparatively large scale. Their leaders then seek to transform the combat from small-scale, hit-and-run combat to regular warfare. So they are tempted to seek the means to strike a mortal blow. Thus, Phase 3 sees the bulk of the fighting.

In his 1937 pamphlet on guerrilla warfare, Mao Tse-tung wrote that in the final phase of the struggle "there must be a gradual change from guerrilla formations to orthodox regimental organization" that can meet the enemy on his own terms. For Mao, timing was crucial: to make the transition before the conditions were propitious was to court disaster. Mao took his time—more than a decade—but some other commanders have tried to move too quickly. In Vietnam, the Viet Minh strategist Vo Nguyen Giap threw his army into three disastrous campaigns against superior French forces during 1951 that almost wiped it out.

Mao's and Giap's transformations of their forces almost certainly were done for objective tactical reasons, but, as I will point out, others appear to have moved for subjective or prestige reasons. Some aspired to match their opponents' formality and even appearance. In the American Revolution, George Washington despised not only the guerrillas who operated successfully far from his command but even the militia on which he had to rely. He found them to be "an exceeding dirty and nasty people [evincing] an unaccountable kind of stupidity." As quickly as he could, he dispensed with both and devoted his energies to creating a British-type army, the Continental Line. As a result, he was defeated time after time and almost lost the war. As his great critic in that war, General Charles Lee, put it, to use "Hyde Park tactics," as Washington wanted to do, was bound to fail. It took years to create the kind of army the British had. The Americans could hope to be only amateurs and to pit them against "a disciplined Enemy, would, in my Opinion, be downright Murder [at best] they will make an Awkward Figure, be laugh'd at as a bad Army by their Enemy, and defeated in every Rencontre which depends on Manoeuvres." But, as we shall see, more leaders followed Washington than Lee. Military commanders choose to overlook the Kenyan proverb that proclaims the power of the flea—"A flea can trouble a lion more than the lion can harm a flea." Most generals would rather be lions than fleas.

Two questions remain. First, what happens if the guerrillas do not get the foreigners and their local allies to give up? And, second, what happens if they do?

The short answer to the first question is that the resistance is almost certain to continue, year after year, even generation after generation.

The second question requires a more complex answer, but to reduce it to the main point, the evidence is that once an insurgency achieves what the leaders and enough of the public regard as an acceptable outcome, usually meaning that the foreigners leave and their local surrogates give up power, the guerrillas become superfluous. At that point, their leaders often become leaders of the government. That is what happened when Éamon de Valera became president of Ireland, Tito became president of Yugoslavia, Ahmad Ben Bella became president of Algeria, and Castro became president of Cuba. At that point, the old guerrilla organization comes to seem not only an anachronism but a threat. The new regime then usually sidelines or suppresses it. That is what de Valera, Tito, Ben Bella, and Castro did. As natives and as national heroes, they could do what foreigners could never do. Thus, the insurgency dies.

With this general background, I will analyze the course of some of the most important and revealing insurgencies and then, in the conclusion, apply the principles—the lessons—to the current American interventions in Afghanistan, Iraq, and Somalia and warn of others being planned.

VIOLENT
POLITICS

CHAPTER 1

THE AMERICAN INSURGENCY

THE BRITISH ARMY WAS PROBABLY THE EIGHTEENTH CENTURY'S MOST highly trained regular military force. Although it was relatively small— on the eve of the Revolution in 1775 it numbered only forty-eight thousand men, about 10 percent of France's military force—it could usually achieve what military men refer to as "theater superiority" because it could be landed virtually anywhere by the Royal Navy. If more troops were needed, it could be augmented by "renting" additional foreign, usually German, armies and by enrolling such natives as Bengali peasants and American farmers. So it was a highly flexible, multinational force with a global reach, but its tactics had evolved in Europe. There, centuries of farming had opened and leveled the terrain. Responding to the landscape, troops were drilled to align themselves in parade-ground formations with each soldier's shoulder next to the shoulders of his two neighbors; they then marched in lockstep to within about twenty paces of a similarly organized foe and, on command, fired what amounted to a broadside. Their smoothbore muskets could not be accurately aimed—they did not even have a rear sight—so soldiers were ordered not to try. Their commanders expected the sheer weight of lead to shock and disrupt their opponents' formation. Neither the British soldiers nor their opponents could reload because of the long bayonets fixed to the barrels of their muskets; so after firing, the soldiers who were still on their feet and able to move charged forward. Their two-foot-long bayonets fixed on their four-foot-long muskets made a virtual spear rather like the ancient Greeks had employed in their phalanxes. These tactics were brutally effective in Europe, where armies routinely slaughtered one another, but made no sense in the American wilderness, where as yet untamed gullies, hillocks, rocks, and trees made rigid formations impossible.

To feed themselves, the British, like other eighteenth-century European armies, pillaged farmers' houses, grain storage barns, and livestock. In Europe, which was relatively densely populated and where, because transport was sporadic, crops were often stored in large quantities, they could do so. But in the sparsely populated and thinly cultivated American hinterland, foraging produced little to eat. So contemporary British soldiers, most of whom were about half a foot smaller than modern soldiers, had to carry huge loads on their backs. Standard packs in the British army weighed fifty pounds, nearly half the weight of the average soldier. Wisely, the French, who were more accustomed to the American wilderness and who moved in smaller formations that made it more likely to be able to live off the land, reduced their loads to about twelve pounds. Indian warriors carried virtually nothing.

Able to move lightly and quickly, the French and their Indian allies refused combat on British terms. As described by Colonel Henry Bouquet, then the immediate superior of George Washington, "they never congregate in a group that could be destroyed by the superior firepower of the regular troops and when attacked they never stand their ground, but immediately give way only to return to the charge when the attack ceases." That is, they fought in the style of guerrillas—loosely, adapting to terrain, quick to attack and quicker to retreat, hanging on the edges of superior forces to pick off stragglers and to disrupt, loot, or destroy the cumbersome supply wagons that accompanied every British force.

Ironically, as we now know, such tactics had been taught to the Indians by the earliest English colonists, who were few and weak while contemporary Indians were many and strong. And, like the eighteenth-century European regular armies, seventeenth-century American Indians regarded the ceremony of war as its essence. War for the British, French, Germans, and Indians could be described as violent theater. They all dressed in gaudy outfits—the British in their famous red coats, the French and Germans in a rainbow of colors and soaring headdresses, and the Indians in a colorful assortment of feathers and furs. They didn't fight so much as paraded before their foes. The Europeans were serenaded by marching bands, playing fifes and drums, and the Indians with choruses of chanters. When they first saw these displays of the Indians, the English colonists hid behind trees and fired into the mass of ceremonial dancers. This was not war as the Indians were accustomed to engaging in it. Heroes could not flaunt their courage, nor could the participants safely display their costumes. The Indians were stunned.

They were not alone. As wars in Europe became more brutal as weapons improved, armies also had to adapt. Cannon cut great swaths through serried ranks, and the soft lead slugs of even primitive muskets made hideous and usually lethal wounds. Gone was the glory and charm of war. Generals tried hard to keep their colorful formations intact but began to supplement them with "detached" troops, who put aside colorful uniforms and fought as what today we would call partisans.

By the middle of the eighteenth century in Europe, *la petite guerre*—the French phrase for the later Spanish word *guerrilla*—came into play. Although regular officers despised *petite guerre* "as not a form of war, but rather a manifestation of criminality," they found it useful. *Petite guerre* troops were used as auxiliaries to regular military formations to gather intelligence, interdict supplies, and harass the opponent. Used only as auxiliary to the main battle formations, they nevertheless sometimes swung the outcome. In the great battle of Fontenoy on May 11, 1745, French partisans managed to delay the advance of the British and allied forces long enough to enable the French army to dig trenches. So when the British advanced shoulder-to-shoulder to almost rock-throwing range of the French, they were halted. As they stood fast, the French opened a murderous fire. Stunned by the weight of lead and prevented by the trench from charging with bayonets, as they were accustomed to do, the British were decimated. Within the first few minutes of the battle, they had lost about one man in each six. Despised or not, the partisans proved their worth.

But the British generals did not like what they saw and moved reluctantly to learn from Fontenoy. General Edward Braddock was one who did not learn. When he set out in 1755 with his colorful, superbly disciplined, and confident British army to drive the French and Indians out of the land west of the Appalachians, he led it into an ambush where his troops were massacred. Of Braddock's 1,100 soldiers, 714 were killed; and of his 86 officers, 63 were either killed or wounded. Braddock never understood what had happened. His force was probably the finest formation in the British army, they outnumbered the French and Indians, their equipment was far better, and they did everything according to the book.

That book was also the bible of George Washington, a young major in the Virginia militia. For him, the British army was the acme of the military profession. He was so impressed by it that he avidly sought a commission as a regular British officer. He was infuriated when, because he was a colonial, the British turned him down. But he never lost his admiration for things British. Later, when he had the opportunity, during the bitter winter he spent at Valley Forge, he desperately tried to whip his ragged, starving colonial recruits into a British-style army, the American "Continental Line." As Charles Royster has written, Washington "wanted a regular, eighteenth-century army that could defend the capital, meet the enemy on the plain, or take New York back by siege and assault." Like Braddock, Washington had failed to absorb the lesson of "informal" or guerrilla tactics.

Washington not only did not understand the lesson; he did not want to accept its implications. The first implication was fundamental to the Revolution. It was that lightly armed, small, highly mobile groups could defeat the superb British army. Benjamin Franklin and a few other Americans saw it more clearly. "The whole transaction," Franklin wrote, "gave us Americans the first suspicion that our exalted ideas of the prowess of British regulars had not been well founded." That may be taken as the first step in the coming war of independence: Braddock had helped convince Americans that they had a chance against the British army. His defeat thus provided a crucial psychological boost to the growing sentiment for a break with Britain.

The second implication was even more deeply disturbing to Washington. It ran counter to everything in which he believed, even during the actual fighting phase of the Revolution. If the world's best professional soldiers could be defeated by the despised Indians, "stiffened" though they were by the French, then what he and the rest of the men of his class almost equally despised, the American part-time militia, might constitute both a military and a political force of truly revolutionary dimensions. As I will illustrate, Washington desperately struggled against both of these lines of thought not only up to the actual outbreak of fighting, but during the entire Revolutionary War. To put Washington's attitude in context, it is necessary to review trends during the years between Braddock's defeat in 1755 and the beginning of the Revolutionary War in 1775. Doing so will take us to the very heart of insurgency.

Although during the 1760s practically no Americans conceived of a "strategy of insurgency" and few (including Washington) even desired a break with England, their daily activities and their commercial interests began to coalesce into such a strategy. Much of what they did came "naturally" to them. They found that colonial governments thwarted their desires for land or were so corrupt that colonists had no chance against the insiders. The laws enacted by colonial legislatures or Parliament, and the implementation of them, proved even worse. The commerce on which colonists depended came to be defined in law as blockade running, customs evasion, and dealing in prohibited items. In many aspects of their lives, they felt annoyed at best and obstructed at worst. As a Philadelphia merchant wrote to Benjamin Franklin, "No one Colony can supply another with wool, or any woolen goods manufactured in it. The number of hatters must be restrained, so that they cannot work up the furs they take at their doors; nay a hat, though manufactured in England cannot be sent for sale out of the Province, much less shipped to any foreign market." Consequently, avoiding or disrupting British administration was both common and popular. Only if the colonists evaded the authorities could they get the goods they needed to live; John Hancock was said to have had about five hundred indictments against him for smuggling, which led his contemporary John Adams to remark that it was the British attempt to curtail his smuggling that made him a patriot. But flouting the law was dangerous; only if they disrupted the British-sponsored colonial governments could the merchants move safely and only if they made impossible the implementation of British laws could they avoid paying the price for disobedience. So, in the decade after the end of the French and Indian wars, what amounted to an unarticulated strategy of avoidance turned increasingly into insurgency.

Collections of colonists gathered themselves into a number of small groups with such colorful names as the Liberty Boys, Mohawk River Indians, Sons of Neptune, and Philadelphia Patriotic Society. Not to mince words, they were terrorist organizations. Working outside the law and at cross purposes with the existing authorities, they tarred and feathered would-be sellers of tax stamps, assaulted customs inspectors, dumped tea into the sea, ran blockades, coerced juries, or prevented courts from sitting. Under their attacks, administrations in the colonies virtually ceased to function. Destruction of the existing administration, as we shall see particularly in the later war in Vietnam, is a stage through which successful insurgencies have to pass.

Massachusetts is the best known of the revolutionary colonies and has been extensively studied, but it was by no means unique. In

New York, such a large mob was led by the town's merchants to the wall of the British-held fort that the royal governor was forced to give up the sale of tax stamps. Even in North Carolina, selling tax stamps was made impossible by the threat or actual outbreak of riots.

Although they do not seem to have had a clear notion of the way the tactics began to constitute a strategy, the rebel groups were driven to struggle not only against the British but also, and even more importantly, against those colonists who sided with the British, the "Loyalists." I think this was an instinctive or visceral act, not guided by any theory or sense of what insurgency requires, but it was crucial in the events leading to the Revolution. Why? It is because, as we now know, America was then not a *society* but a collection of unrelated groups. Leave aside the obvious racial, ethnic, religious, and linguistic partitions that were nearly impermeable and look just at the political divisions: perhaps six and almost certainly at least five in each ten white Americans were more or less content with the status quo or, as the American historian John Shy has written, "were dubious, afraid, uncertain, indecisive, many of whom felt that there was nothing at stake that could justify involving themselves and their families in extreme hazard and suffering." Those of English background who lived along the Atlantic coast were often in very close contact with England, to which they sold many of their products and from which they bought almost everything from toiletries to tools. They formed their opinions from reading the English press and often sent their sons to study in England. Those of German background lived in virtual apartheid, managing their own affairs, and were little interested in the growing disagreements among the English speakers. And most of the people who had pushed into the interior, often Scotch-Irish, were too engaged in clearing their lands, getting enough to eat, and keeping the Indians at bay—and too out of touch with events along the seacoast—to pay attention to agitators in Boston or Philadelphia. Such contented or otherwise occupied people had to be won over to the cause of the revolution or cowed into silence.

Of the remaining four or five in each ten, two were Loyalists. They, more than the British, were the really dangerous enemy of the would-be rebels. Without them, the British could not maintain their rule over the colonies. So, all over colonial America, the insurgents engaged in violent actions to terrify, drive away, or kill the Loyalists. They succeeded. By 1775, as support for the insurgency swelled, tens of thousands of Loyalists had fled to Canada.

What is most striking about these events was that there was no organization or shared set of beliefs beyond a single colony or even in most cases beyond a town or settlement. It was what might be called the "local logic" that drove similar actions among groups scattered over the great distances that separated the colonists from one another. There was no party structure, as we shall see in Vietnam, for example; no idealized shared memory, as we shall see in Ireland; no ideology, as we shall see in Yugoslavia; no sense of racial, ethnic, or religious animosity, as we shall see in Kenya; and above all, at least at the outset, no clear sense of nationalism as is common to most insurgencies. But the record shows that all over America little groups, with no discernible links to one another, began to engage in small, local acts that were astonishingly similar.

Most of these informal, small-scale, and frequently ugly actions have slipped our collective memory. Americans today would rather concentrate on the glorious episodes, but the numerous small events played at least as important a part in the American insurgency and the break with England as the beloved battle of Bunker Hill. Because they are now obscure, I will mention just enough to show what they amounted to, focusing on what I call the climate of insurgency.

That climate, the sense of alienation from the existing political order, grew mainly from two sets of British policies. The first of these was the British attempt to calm the frontier and prevent further Indian uprisings, like the costly resistance of the great Indian guerrilla leader Pontiac in 1763. In an effort to stop the drain on their treasury by constant warfare, the British realized that they had to prevent white settlers from invading Indian territory, so they issued a proclamation to ban movement beyond the crest of the Allegheny mountains. This policy alienated all those colonists who had come to America in quest for land. In their thousands, they refused to abide by the Crown ordinance and continued to trek westward. Not only the actual settlers, usually poor, ill-educated, scattered, and despised by both the British and the more established colonists as nothing more than "white Indians," but richer, more powerful speculators and even Crown officials were converted into subversives by this policy. They too wanted to profit from free land. The final British move against them came in 1774, on the eve of the Revolution, when the British government made a vain effort to void all western land grants, thereby infuriating the growing number of frontiersmen who, having chopped down trees, hoed and plowed plots of land, planted and harvested, built shacks, and begot children, firmly believed that they had turned their "tomahawk claims" into actual rights.

The second set of British policies had a similar effect on the growing population of urban merchants and farmers along the coast. Alarmed that the colonists were subverting British rule, insulting, harassing, or even imprisoning Crown officials and refusing to abide by parliamentary laws, the British cracked down in a series of actions and new laws. These culminated in the promulgation in June 1774 of what the insurgents called the "Coercive Acts" or "Intolerable Acts." They were essentially intended to punish the recalcitrant colonists, particularly those in Boston. The Boston Port Act effectively closed Boston to commerce and was a death sentence on the merchant community of that city; the Massachusetts Government Act enlarged the powers of the royal governor at the expense of the already traditional legislature; the Administration of Justice Act provided for trial without jury in a "vice admiralty" court for certain offenses that customarily had been handled in the colony and authorized the rendition of people charged with sedition to England; the Quartering Act allowed the British effectively to seize property from the colonists; and, finally, the Quebec Act stripped Massachusetts, New York, Connecticut, and Virginia of lands they claimed beyond the Alleghenies and awarded them to Quebec, whose Catholic status was affirmed. This last act thus raised in the colonial context the very religious issue that had caused many of the colonists to leave England.

These actions by the British government drew a sharp contrast between what the colonists had come to regard as their rights and what they saw as illegal British innovation. The rhetoric of colonial agitators was now proclaimed in bold script by the British government. Taken in sum, the Coercive Acts offended almost every sector of the colonial society: merchants lost trade; settlers lost at least the dream of cheap land; householders feared confiscation of their property; the religious members of the population, who were almost entirely Protestants, saw the dread hand of the papacy, which they had escaped in the Old World, thrusting toward them in the New; and those who were testing the boundaries of mercantilist regulations or political prohibitions feared that they would be dragged off in chains to British courts where they had no protection. Thus, these acts virtually forced a climate of insurgency upon the colonists. For the first time, large numbers of colonists began to share the sense of opposition to the British Parliament.

At the same time, however, the acts split the colonists. An equal number either continued to feel a sense of loyalty to the Crown or feared the turmoil that accompanied the new political awakening of their neighbors. What began as little more than a gentlemanly or even theoretical debate inevitably turned to ugly and violent clashes between Loyalists and those who were beginning to think of themselves as insurgents. As the American historian John Pancake has written, "the real prize of the American Revolution was the allegiance of a majority of the people who, in the early stages of the war, did not commit themselves. The winning of independence was, in the final analysis a numbers game. The objectives were not so much New York or Charleston or the Hudson River Valley but control of the civilian population." It was population, not land, that, as Mao Tse-tung later wrote, determined success in insurgency.

What Mao Tse-tung famously called the "sea," that is the citizenry, coalesced into a body that was prepared to support armed opposition to the British. As they did, the militants who sought to shake the civic order, the people Mao called the "fish," began to act. As we shall see in later insurgencies, the two groups were caught up in a mutually reinforcing process: as violent acts were committed, the public was drawn in, and as it became more supportive, violence increased. Much of this was so undirected as to appear almost automatic. In colonial America, there was no central organization or guiding ideology that would be evident in later insurgencies. Even the catch-all concept of nationalism was still embryonic, but as John Shay has written, the early, and still informal, combat, particularly between the Loyalists and the Rebels, "was a political education conducted by military means . . ."

Few American historians have treated the rebels as guerrillas in the context of the climate of insurgency during the Revolution. Neither did contemporaries. The concept of *guerrilla* as a popular resistance through political, administrative, and military means was not defined or perceived until the Spaniards began to fight against Napoleon's forces nearly half a century later. Yet, in the first stages of the American rebellion, some of the combatants were organized and fought in the manner of guerrillas. American historians refer to them as "Irregulars." But they were different from what the Europeans thought of as irregular or auxiliary troops.

What differed in the American Revolutionary War was that, for the first several years of the war, particularly in the southern colonies, there was no regular colonial army to which "auxiliaries" could be attached and no formal command and control structure that could mobilize and use them. When the British overran South Carolina, the rebel government collapsed. Then the men who had already proven their ability as partisans were on their own. They had to become political as well as military leaders. Elijah Clarke, Thomas Sumter, Francis Marion, William Davie, and Andrew Pickens gathered men and arms on their own authority and fought on. As an early biographer of Andrew Jackson, who served with one of them, wrote, they "were for a time independent chiefs, receiving neither pay, subsistence, arms, nor stores from government, but fed and clothed by the contributions of friends and the spoils of enemies, armed with such weapons as could be found or made in the country." We will hear similar words spoken of Tito's guerrillas during World War II.

Also different from the European concept of *la petite guerre,* whose practitioners were often mercenaries and nearly always foreigners, the American rebels worked within a social context. As the British general Thomas Gage wrote, the popular mood was "rage and enthousiasm [sic]." Already by 1775, all over colonial America, cities, towns, and tiny settlements were forming groups in opposition to British measures or to Britain itself and were beginning to assemble means of resistance and training themselves to effect it. Their first significant engagement in the south was the battle of Moore's Creek Bridge on February 27, 1776. In that engagement, a hastily gathered group of farmers and townsmen-turned-insurgents headed off the attempt of some fifteen hundred Scots Loyalists to join the British army near Wilmington. The battle was dramatic but local.

More widespread and significant than the military engagement, on which most historians have focused, was that local organizations, usually calling themselves Committees of Safety, began to purge their communities of Loyalists, sometimes with propaganda but, as they increased in number and power, often with terrorism. As Catherine S. Crary writes, in their eyes, "every Tory became a potential informer or saboteur, a fifth columnist. The Committees of Safety clamped down more strictly than ever on the moderates and dissemblers who heretofore had been left alone . . . Washington, who in 1775 had stipulated that a person would be considered a Tory only when he had given 'pregnant proof' of an 'unfriendly disposition to the cause,' now called them 'parricides' and 'unhappy wretches! Deluded mortals!' 'A great many ought to have . . . long ago committed suicide.' Major General Charles Lee spoke of the Tories on Long Island as 'dangerous banditti' [a phrase we will often hear in the future conflicts], adding 'not to crush these serpents before their rattles are grown would be ruinous.' Animosities were now afire."

So effective were the Committees of Safety that the Loyalists, even when led by British officers and furnished with British arms, were rarely able to organize counterinsurgency groups. As John Pancake has written, "by the end of 1777 the British generals had discovered that Loyalists, if they existed in large numbers, were not always willing to declare themselves, much less volunteer for soldiering.

Burgoyne looked for them in the Hampshire Grants, in the Mohawk Valley, and on the upper Hudson—and he looked in vain. Howe's expedition to Pennsylvania was grounded in his conviction that the presence of large numbers of Loyalists would ensure the overthrow of Whig rule once Philadelphia was taken. When he landed at the Head of the Elk [river in Maryland] in the summer of 1777, he was greeted by a handful of Loyalists and by columns of smoke on the horizon marking the fires set by Patriot farmers burning their crops."

Following engagements, insurgents often summarily executed captured Loyalists. After the battle of Kettle Creek on February 14, 1779, for example, they judged sixty-five of the men who were fighting on the British side guilty of treason and immediately hanged five. Terrorism had moved to a new stage; it had become quasi-judicial: among the first acts of the newly empowered legislatures were laws dealing with Loyalists. Treason was then defined as helping the British in almost any way, so that many people were imprisoned, driven away, or executed just for making comments in public that were read as "Tory."

More important, suppression of the Loyalists had become a tactic in the military struggle. With their armies stretched across hostile countryside, the British were unable to protect the Loyalists they encouraged to come forth. If they tried to do so, as General Cornwallis soon found, protecting their would-be auxiliaries drained the manpower of his regular forces rather than augmenting it. As he wrote on August 6, 1780, "The whole Country between Pedee & Santee [rivers in the Carolinas] has ever since been in an absolute State of Rebellion; every friend of the Government has been carried off, and his Plantation destroyed; & detachments of the enemy have appeared on the Santee, and threatened our Stores, & Convoys on that river." We shall see this pattern repeated in the way the French were stretched out in Vietnam.

Threat was also used as a recruiting incentive. One Scottish visitor describes in July 1775 in North Carolina what probably was a common occurrence. "An officer or committeeman enters a plantation with his posse. The Alternative is proposed: Agree to join us, and your persons and properties are safe . . . if you refuse, we are directly to cut up your corn, shoot your pigs, burn your houses, seize your Negroes and perhaps tar and feather yourself."

Tarring and feathering was not an idle threat. No one knows how often mobs tarred their victims, but enough reports exist to indicate that it was common. As the hot tar burned the skin it was painful; worse, it was humiliating since the victims, naked except for the feathers and made to look ridiculous, were usually paraded through their towns to be pelted with mud and excrement by crowds of children and cursed by adults. Far more painful and also apparently common were lashings administered to suspected Tories and especially to those suspected of supplying the British troops with information. In some instances, the lashings were followed by threats like one given a young boy that he would be staked out on a beach and allowed to drown in the incoming tide. Those men who were captured either as a result of denunciations or on suspicion were often thrown into makeshift jails that rivaled the worst of European dungeons. Many were lynched by mobs or ordered by ad hoc Committees of Safety to be hanged.

What the American insurgents were doing piecemeal foreshadowed a pattern. Everywhere insurgents spend more energy attacking their fellow inhabitants than the foreign enemy. They do so in part, no doubt, because civilians are easier targets than soldiers, but this is not the crucial reason: it is that unless they can forge a solid core of like-minded people, they cannot hope to survive, much less to "win." Those natives who join the foreigners are more dangerous to the national cause than the foreigners themselves because, unlike the foreigners, they have the capability to form a native government. Thus, in America in the 1760s and early 1770s, as in each subsequent guerrilla war including most recently Iraq, insurgents more often attack their recalcitrant fellows than the foreign soldiers. By war's end, nearly one hundred thousand—roughly one in each twenty colonists—had fled.

As it became necessary to fight the British as well as the Loyalists, the inexperienced American officers were at a loss to know what to do. As I have said, George Washington's answer was to form a British-style army. What he managed to put together after Valley Forge could not match the British in battle. With few exceptions, Washington had to avoid battle.

Only one senior commander, Charles Lee, seems to have understood the political or social dimension of insurgency. For him it was manifest in the very issue of military tactics. He despised the formal military notions that Washington accepted so readily from the British as "Hyde Park tactics." Rather, Lee placed major emphasis on securing the population for the rebel cause. As John Shy writes, in the early months of the war, "people on the American seaboard outside Boston were wavering. There had been little fighting and many people, caught up in the shadowy state between war and peace and loyalty and treason, made small compromises with British officials. There seemed to be a tendency to listen more closely to loyalist neighbors now that an abstract argument had become immediate and concrete, and to postpone any violent or decisive action as long as possible. Lee moved through this murky world of indecision like a flame." He forced the undecided to take oaths of allegiance to the revolutionary authorities and made those authorities sever all connections with the royal governments. That is, he took the first and most important step in insurgency, securing the political element of guerrilla warfare. He then immobilized all those who were not won over politically: he regrouped, as we would say today, suspected Tories, moving them to remote areas where they could not aid the British or the Loyalists, anticipating what the Americans would do in Vietnam with the strategic hamlet program.

Lee also anticipated another tactic adopted by insurgents in Yugoslavia, Greece, the Philippines, and Vietnam: he sought to create more or less permanently held rebel-controlled areas in which they would administer the services of government, provide sanctuary for troops, produce food and other supplies, and serve as recruitment depots. Such mini-states were what the French called "oil spots" (*tache d'huile*). The French used them in their struggle with insurgents in Madagascar and Vietnam as outposts for their army of occupation, but the Viet Minh insurgents used them in an opposite sense, indeed exactly like the Americans, spreading the "spots" they controlled over the countryside. Many Americans never saw British troops, but even those that were on the British routes of march quickly reorganized once the British passed through.

Lee realized that the British could not hold everything, so it was logical to deny them even the results of their frequent victories. The way to accomplish this was not to rely upon the regular forces—the Continental Line on which Washington had placed his emphasis— but on the local men who knew the ground and were associated with the inhabitants. They would provide the fist that enforced local regulations and public order since they had reason to protect their property. They would be much more effective, he argued, than the British-style army Washington wanted (and had Friedrich Wilhelm von Steuben teach his troops at Valley Forge). As he put it, "If the Americans are servilely kept to the European Plan [of tactics, which aligned troops as though on parade] they will make an Awkward Figure, be laugh'd at as a bad Army by their Enemy, and defeated in every Rencontre which depends on Manoeuvres." From personal experience, Lee knew that the Americans would not have time to create European-style soldiers. If they tried, as another American officer, George Washington's kinsman Robert Washington, put it bluntly, "to lead a Body of brave men, with such counterfeit Discipline, to face a disciplined Enemy, would, in my Opinion, be downright Murder." What the Americans should do, Lee counseled, was to recognize their own skills and strengths rather than trying to ape those of the British. That is, working from areas where the British would have trouble operating and sallying out from these areas "harassing and impeding [which] can alone Succeed."

The British certainly misread the American insurgents' actions and added them to the other obvious signs of disunity and weakness in the colonies. They simply could not believe that poor, isolated, ill-equipped, ill-trained, and mutually destructive American colonists would even dream of fighting the mighty British Empire. The favored dismissal of insurgents was that they were just "bandits," as

Napoleon would refer to the Spanish guerrillas; British prime minister David Lloyd George, to the Irish Republican Army (IRA); German officers, to the Yugoslav Partisans; and American military men, to the Greek ELAS.

When it came to the colonial militias, British scorn was even greater. A senior British general told Parliament on February 2, 1775, that the Americans could not fight and that with just five thousand soldiers he could reestablish British rule. General James Wolfe wrote that the American soldiers were "the dirtiest most contemptible cowardly dogs that you can conceive. There is no depending on them in action. They fall down dead in their own dirt and desert by battalions, officers and all." Another British officer, General James Murray, thought he knew why. As he put it, "the native American is a very effeminate thing, very unfit for and very impatient of war."

George Washington agreed with their assessment. For him, the common Americans were "an exceeding dirty and nasty people [evincing] an unaccountable kind of stupidity in the lower class of these people." When in June 1775 he arrived in Cambridge to take command of the Revolutionary army, he found that he did not have one. It was just a collection of town militias eager to go home. Washington had no money, no staff, no commissary, few serviceable arms, even fewer men who knew how to shoot, and little ammunition. Nowhere in the colonies was gunpowder then produced—it had always been imported from England—and no domestic source of flints had yet been found. Without flint, the guns of the period could not be fired. About one soldier in each four had no weapons of any description. Many did not even have serviceable shoes. Not surprisingly, George Washington commented, "When I took command of the army, I abhorred the idea of independence." One of Britain's most able officers found the American pretensions "a mystery indeed."

The solution to the mystery was a sequence of actions and attitudes that fell into a clear pattern: they began with the perception that the enemy was vulnerable. From that point, terrorists immobilized the British administration and began to coalesce, invigorate, or purge the general public. Next, they established their own administration in mini-states that gradually spread across colonial America. From these safe havens, they began, slowly and on a small scale at first, hit-and-run guerrilla warfare to wear down and dispirit British troops. What they accomplished in the Carolinas and Georgia prepared the way for the campaign that chased General Cornwallis to Yorktown, where the British surrendered.

But already in the American Revolution, and in later insurgencies, we see the tendency of guerrilla leaders to give up their successful tactics and attempt to act like regular troops. When the means became available, that is, when a regular army, the Continental Line, was created, the groups that had been guerrillas faded rapidly into the role of auxiliaries or themselves became militia units. They did this, in part, no doubt because they or their leaders found it more prestigious to be considered soldiers than guerrillas; they were also less likely to be hanged by the British if captured while wearing some sort of uniform. Perhaps even more important was the fact that the new American military leaders tried to duplicate the British order of battle and despised the ragtag guerrilla gangs. George Washington's diaries are full of deprecating references to them, their lack of discipline, their inattention to soldierly deportment, and their inability to face up to British formations. His disciple, General Nathanael Greene, wrote that "Partisan strokes in war are like the garnish of a table," which "afford no substantial national security."

Certainly Greene was right that the American guerrillas offered no substantial national security. No guerrilla force does. But groups formed by military leaders like Francis Marion, Elijah Clarke, Thomas Sumter, and Andrew Pickens disrupted the plans of the British forces, intimidated their Loyalist allies, captured or destroyed wagon trains of supplies, and encouraged popular resistance. Like many later military leaders, George Washington did not understand that process. He so despised the people who engaged the British in this ungentlemanly fashion, and was so anxious to sideline them, that he came close to losing the Revolution. He might have actually done so except for the July 1780 intervention of six thousand French soldiers and the September 1781 French naval blockade of

Cornwallis's army at Yorktown. The French saved the Revolution. And, together with Washington's Continental Line, they did so in a conventional military way.

The story does not end there. As in many insurgencies, there was a sequel. After the British and French withdrew and the Continental Line stood down, the men who had made the "little war," and whom Washington had sidelined, continued the revolution in other forms. Indeed, in a sense, the political movement of Andrew Jackson half a century later was an echo of this earlier insurgency. This "after-war" would be repeated in other wars in Algeria, Afghanistan, Ireland, and the Philippines. So let us pay close attention to those who become the soldiers of the little wars. For this I turn to the episode that gave this form of insurgency its name, the Spanish *guerrilla* against the French.

CHAPTER 2

THE SPANISH *GUERRILLA* AGAINST THE FRENCH

DESPITE HAVING CONQUERED A LARGE PART OF THE KNOWN WORLD in the sixteenth and seventeenth centuries, Spain remained a weak state. The vast treasures it looted in the Americas paused only briefly in Spain before being paid out to England, France, and the Low Countries for luxuries the Spanish court and privileged class craved. Inca gold made almost no impact upon the lives of most of the population. At the end of the eighteenth century, Spanish society struck the few visitors from other countries as poor, illiterate, and even brutish. As we would today describe their impressions, they found Spain still living in the Middle Ages. Worse, they were not progressing but receding, with urban populations dwindling. As Napoleon taunted them in a proclamation of June 3, 1808, "Spaniards: remember what your fathers were like and look at what you have become." With only the most primitive means of communication and transportation, Spain was not so much a nation as a collection of autarkic villages and towns, each making its own tools and clothing, growing and consuming local crops, and even speaking distinct dialects. The major unifying institutions were embedded in the Church.

The Spanish Church permeated every hamlet. French visitors described Spain as "one vast temple." As Raymond Rudforff wrote, "No other Christian country, not even Italy, had so many religious institutions, monuments and shrines. A quarter of all the land which could be cultivated was Church property. Nearly 200,000 of Spain's ten million inhabitants were either members of the Church or else directly dependent upon it. There were some forty different orders of monks, nearly thirty nuns' orders, over two thousand monasteries and more than eleven hundred convents. The numbers of the clergy, the respect they obviously enjoyed at all social levels, their wealth and their influence convinced foreigners that Spain was above all a nation of monks and priests." The Church did not rely just on numbers. The Inquisition was still a feared institution whose propaganda pervaded every Spanish household. But loyalty to the Church did not rest on fear alone: gluing together the structure of Spanish society despite its schisms and regional differences was a deep-seated and vigorous religious fervor.

Its monarchy also unified Spain, despite the fact that by the early years of the nineteenth century, it was but a pale remnant of its great tradition. The king and his court became the butt of jokes by later historians: the Bourbon king Carlos IV was a simpleton; his wife, Maria Luisa, the granddaughter of Louis XV, was the mistress of a commoner. The ambitious crown prince, Fernando, was stupid enough to be caught inciting sedition against his father. Courtiers aped the styles of pre-Revolutionary France—except that the women were veiled and the men prided themselves on their ignorance of modern thought. However, what happened at the court was largely unknown to the public. For the common people, the monarchy was still the grand institution of the Habsburg king Karl or Carlos V. It was their pride, and to it they were blindly loyal; consequently, they would make the weak and vacillating Fernando the symbol of their resistance to the French.

On such practical matters as taxation and control of commerce, however, loyalty to the Crown was balanced against provincial autonomy. The medieval states that merged into Spain kept their separate institutions. *Fueros* that enshrined the terms on which they had merged with Castile and Aragon were jealously guarded proclamations of local rights. Attempting to abolish them and to centralize Spain, as the French had centralized their own country after the Revolution, would be one of the most provocative of the French policies.

Relations with the French before the invasion of 1808 were crucial to the functioning of Spanish society and to its foreign affairs. Such industry as existed in the cities was primarily worked by Frenchmen whom the Spaniards deprecated as *gavachos* (a word that had more or less the sense of the American "wetback"). In the northern cities, they numbered in the tens of thousands. About one in each five inhabitants of Madrid was reputedly of French origin, and in the generation before the French invasion the census listed twenty-seven thousand "heads of family" as French. Even in the countryside, where nine in each ten Spaniards lived, the French were everywhere. As Fernand Braudel has written, "Spanish land would often have remained uncultivated without the peasant who came from France." Perhaps from their role in trade and their competition in agriculture, the French were generally disliked. That dislike intensified after the French in their Revolution of 1789 overthrew their own monarchy and attacked the Catholic Church.

These were the crucial elements affecting Spanish government and society when Napoleon sent his great cavalry leader, Marshal Joachim Murat, into Spain with one hundred thousand troops in 1808. There was no question that France's army was then the greatest in the world. So awesome was its power—and reputation—that the Spanish army withdrew to its barracks in Madrid and declined to fight. For their prudence, Murat publicly, and humiliatingly, thanked them.

No braver than the army, the monarchy temporized. Or, more accurately, it split apart. Crown Prince Fernando wrote a groveling letter to Napoleon charging that his father had been misled into an anti-French policy by "a small clique of malevolents" and containing the seditious suggestion—for which Napoleon later taunted him— that he be given a "princess of your noble line" as a wife. Furious at the discovery of his son's "reckless and outrageous plot," the king arrested him and humiliatingly confessed to the general public that he had been unable even to control his own household. The prince was then forced to issue a statement begging permission "to kiss your royal feet."

With these domestic troubles besetting him and seeing that effectively he had no army, the king felt constrained to smooth the way for Napoleon's army, telling his people that "the army of our beloved ally, the Emperor of the French, is crossing our kingdom [to attack the Portuguese and their British allies] in a spirit of peace and friendship." Although he acknowledged that the French were actually planning to stay in Spain, they would do so, he said, only "in those places which lie under threat of enemy [that is, English] landings." The proclamation failed to calm the population or to prevent the people from demonstrating their hatred of the French. The king tried to stand aside by retiring to the summer palace at the little town of Aranjuez. That move gave Fernando's supporters their opportunity. They used it to stage a palace coup and force the king to abdicate. While the coup was bloodless and on a small scale, it caused Napoleon to comment to the would-be king Fernando that "it is very dangerous for kings to allow their subjects to get into the way of spilling blood by taking the law into their own hands." It was a portent of the future guerrilla war.

Ever the grand strategist, Napoleon was not to be sidetracked by petty monarchs. With Spain under virtual occupation, he believed that all power was in his hands. Fed up with the Spanish royal family and feeling that he no longer needed them, he decided to exile them. When he did, to his surprise, demonstrations broke out all across Spain in acclamation of the crown prince, whom Napoleon had already come to despise. So he summoned the erstwhile king and the would-be king to meet with him. The king was willing to do anything Napoleon wanted, and to the crown prince Napoleon made a mafia-like offer. According to a contemporary Spanish writer, he said to Fernando: "It is necessary to choose between abdication and death." Fernando quickly saw the logic. He abdicated. Napoleon had already decided to make his brother Louis king of Spain. So sure was Napoleon that his overwhelming military power meant that the war in Spain was a mission accomplished that he completely neglected Spanish xenophobia. His was a mistake that many future invaders would make. Had he kept at least the façade of a Spanish monarchy, he might have avoided the war to come.

At first, Napoleon seemed to have calculated correctly. Such bureaucracy as Spain had shifted easily to the new order. So did the leaders of the Church. The French army's shock and awe had convinced the Spanish army that resistance was useless. The troops who might have resisted, the best equipped and trained, Napoleon wisely dispatched to France to be later wasted in his Russian campaign. So all Spanish institutions that mattered were realigned in the new order. But then things began to go wrong.

Two months after the invasion, a popular insurrection broke out in Madrid. With the army either exiled, immobilized, or defeated and most of the established institutions of the state co-opted, the common people were angry but leaderless. So their initial revolt was bloody but not threatening. As they would do later in Algeria and Vietnam, the French moved to suppress it brutally. Still remembering the Reconquesta that had driven out the Muslims, the Spaniards were particularly shocked that the French used Egyptian Mamluk cavalry against them. Francisco Goya made them the subject of one of his most famous paintings. The next day, the French proclaimed that anyone carrying arms or passing out libelous pamphlets was to be shot, any gathering of more than eight persons was to be fired upon, and any place where a Frenchman had been killed was to be burned down. The bloody suppression of the uprising was immortalized by Goya in a painting that became the icon of the Spanish war, *The Third of May*.

Goya captured the mood of the Spanish population. The Second of May insurrection and the Third of May retaliation constituted the sparks that fired the resistance. Volunteer horsemen took copies of a nongovernmental proclamation urging resistance to the French to towns and villages all over Spain. In response, and often with encouragement from the Church, the common people brushed aside the existing government officials and formed *juntas* (committees) to organize resistance. Most of the *juntas* were socially conservative, indeed, monarchist—proclaiming Fernando their king—but their members were often poor and many were even illiterate. They had no program except for hostility to the French, and they would have been shocked to have been considered revolutionary in a social sense. Wherever they could, they co-opted aristocrats or royalist generals to be their leaders.

Many of the towns that proclaimed their opposition were too small or remote to demand French attention, but one city the French could not overlook was Saragossa because it was a junction on the supply route from the Mediterranean coast to Madrid. Moreover, its challenge to the French was serious. Just three weeks after the Second of May insurrection in Madrid, university students, members of the craft guilds, and peasants from the surrounding farms rampaged through the city, overturning its pro-French or "loyalist" government and demanding that the French leave Spain. They armed themselves with some twenty-five thousand muskets and sixty-five cannon they found in the local armory, and installed a wealthy local nobleman who was also an officer in the royal bodyguard as leader of their *junta*. Under his direction, the city transformed itself into a military camp. Within days, about ten thousand men were being trained how to shoot.

Rising to the challenge, the French sent a flying column of experienced infantry toward the city, driving the amateur Spanish troops before them. That, as the Americans would find later in Iraq, was the easy part. Then, after bursting through the city walls, they found themselves engaged in a different sort of war. In Saragossa, they were trapped by the labyrinthlike medieval city streets. There it was the townsmen, not their new army, that came to grips with the French. Unaccustomed to street fighting, where even women and children could be lethal, the French lost some seven hundred men. They pulled back temporarily. Meanwhile, other French forces assaulted dozens of smaller towns and even the great city of Cordova. The sack of Cordova showed the people of Saragossa and other cities what a French attack involved: the city was looted, men were massacred, and women were molested; by ransacking the cathedral and gang-raping nuns, the French made the Church their enemy, and it responded in its traditional way. Miracles proclaimed the defeat of the invader. Priests exhorted the people to take up arms and fight the "Godless French." Foreshadowing the role of mullahs and imams in modern Islamic resistance movements, priests proclaimed, "Heaven will be attained by killing the French heretical dogs" and promised that "any soldier wounded fighting the French was ensured 100 years relief from purgatory [and that] anyone killed would be reborn three days later in paradise."

Returning to besiege Saragossa, Napoleon's generals wanted to leave no doubt that the Spaniards were defeated. Militarily they were. A destroyed Saragossa surrendered. But the French victory there was self-defeating. As the Spaniards interpreted the French "victory," it was a declaration of war, not a proclamation of its end. Saragossa became Spain's Stalingrad. Its desolation fed popular fury. The battle for Spain became a people's war. Unlike the great battles Napoleon had fought elsewhere in Europe, in Spain there could be no quarter, no victory, no end. As Americans would come later to view their war in Vietnam, Spain became a quagmire into which men and treasure were poured only to drown: during his later exile on Saint Helena island, Napoleon would write, "That unfortunate war destroyed me."

Meanwhile, as events unfolded, the Spaniards were painfully learning that they could not defeat French armies in battle. But, at the same time, the French were learning that to defeat the Spaniards, they had to destroy the country. Both sides found the cost unacceptably high. The French settled down to rule what they could, the cities and key transportation links among them, while the Spaniards, no longer trying to match the French armies in the field, were forced into a new kind of warfare, the *guerrilla*.

At first, the French could not understand this new kind of war-fare. As I have pointed out in the previous chapter, they were famil-iar with what was known among strategists as *la petite guerre.* The Spanish word means the same as the French phrase, but the forms of conflict were entirely different. In *la petite guerre,* detached, irregular, and often foreign soldiers operated as auxiliaries to regular forma-tions; in *guerrilla* there were no regular forces to which the *guer-rilleros* could attach themselves. The *guerrilleros* were local people, amateurs who carried on with their regular means of earning a living, who fought when they saw an advantage and faded away when they were outgunned. A contemporary German officer described the di-lemma of regular armies' combat against them as like "the fabled lion who was tortured to death by a flea."

Rural insecurity was disturbing, but initially the French did not consider it serious; it amounted just to hit-and-run attacks being carried on by small groups of ill-armed peasants, most of whom, the French believed, were just bandits. Cowards who ran at the first sight of a French patrol, they had to be driven into combat by hotheads, *les meneurs,* who did not even qualify as "remnants" of the former regime. They had no claim on any sort of political legitimacy. The French thought of them as the British viewed the American insur-gents: a rabble. But no matter how hard the French struck them, they did not go away.

Spanish geography lent itself to this new form of warfare. When they first marched across Spain to invade Portugal in 1807, the French found their route a daily torture. Unlike the lush farmlands of Germany, where soldiers could always loot a barn, the land was a barren, windswept wilderness punctuated by mountains in which scattered collections of huts announced a nearly starving population. Particularly in the north near the French frontier, in the province of Navarre, where most of the guerrilla actions were to take place, Spain was "a land of dramatic precipices, gorges, hidden valleys, and box canyons with their secret exits." The countryside was virtually impassable for French cavalry and artillery. Village by village, val-ley by valley, bands of desperate men suddenly appeared to ambush and then just as rapidly to run before troops could gather. When they could, they overwhelmed isolated outposts and wagon trains. Always, they carried off their loot to hideaways beyond French recovery. The very isolation of the area and dispersal of its inhabitants, both factors that had prevented them from organizing a powerful national army, now became their greatest asset. Indeed, so scattered were the Spaniards that the French could rarely even find them.

Geography favored them. But the people of rural Spain were also driven by a code of honor that demanded manliness and revenge for offense. The society of Navarre was especially active in the resistance. The north, known as the Montaña, was agricultural and divided into small family plots. There were no large owners and few landless laborers. The people were poor and were too numerous for their land. Consequently, they had long supplemented their meager income by hunting and smuggling goods to and from France. Every man carried arms. Often operating outside the law, they were skilled in the byways of their mountains and valleys. Proud and independent, they were accustomed to operating on their own or in small bands (known as *partidas*) with their kinsmen, neighbors, or friends.

Apparently they began to turn to guerrilla warfare because the French forces, running beyond their supply lines, began to requisition supplies from them. That was the customary means of supply of armies all over Europe, but the Spanish villagers were so poor that such scavenging meant starvation. So the villagers took what they could carry and ran away. Angry and hungry, the French would burn the village in reprisal. Also angry and hungry, the peasants would retaliate.

Not many men were involved in each small episode. One of the most famous of the guerrilla leaders, Javier Mina, had only twelve men with him in his first engagement. He ambushed ten French soldiers and took their arms. Emboldened by his success and having better equipment, he gathered a few more villagers, and within a month, his *corso terrestre* ("dry land pirates"), as such guerrilla bands were called, numbered about two hundred. Each raid netted more equipment and attracted more followers. One was recorded to have overtaken a convoy conveying prisoners to France; when liberated, many joined his band.

As they raided French convoys and outposts, the guerrillas began to acquire valuable loot that they divided among themselves. When they fell short or wanted more, they levied contributions on villages and, drawing on their earlier experience, organized smuggling and even took over the control of the royal customs on the French frontier. Eventually, as the scale of their raids grew and as targets were picked with more care, the bands acquired sufficient resources to pay the combatants regular salaries. Such a flow of cash and goods, in turn, encouraged their supporters to help them in a variety of ways. One was the supply of information, for which they paid handsomely. They were able to buy clothing and smuggle it out literally under the noses of the French forces occupying Pamplona—the contraband clothing was hidden under caskets by a local undertaker. And, most important, they were able to repair the muskets they captured from the French and even to manufacture shot, powder, and guns. (As we shall see, they would be followed in creating a small-scale arms industry by the movements led by Tito, Mao Tse-tung, and Ho Chi Minh.)

How many were these "guerrillas"? The Duke of Wellington, who won the battle of Vitoria partly because the guerrilla force of Javier Mina diverted nearly twenty thousand French troops, deprecatingly remarked that they were just "a few rogues." He was denying the reports of his own intelligence service, which in 1811 put the number of guerrillas at about twenty-eight thousand operating in some 111 bands. Adding those to guerrillas in other parts of Spain, René Chartrand comes up with an estimate of about thirty-five thousand combatants in total. As in all guerrilla wars, the numbers varied as men returned to sow and harvest or to sally forth to attack inviting targets. But even the smaller number suggests a total, including their active supporters, suppliers, and informants, of perhaps a quarter of a million people and an addition of "passive" adherents of several times that many. So perhaps one out of every four adult Spaniards was in one way or another associated in the anti-French movement.

With such resources, they were able to blockade cities as large as Pamplona and ultimately by March 1813 to make French control of Madrid impossible. Given that the French forces reached about four hundred thousand at the height of their campaign and must have averaged at least half that number, how did the Spaniards do it?

The guerrilla strategy, which evolved in the circumstances of the confrontation, came down to two key activities. The first was to eliminate or neutralize those Spaniards who tried to collaborate with the French. This was essentially the same as the American insurgents did with the American Loyalists. Within a year or so after the fight began in 1808, those who sided with the French were safe only in cities garrisoned by French troops. Later, when the French evacuated Madrid, they took with them these *afrancesados*. The second aspect of guerrilla strategy was to draw the French out to protect their supply lines. The French answer, then and later in Vietnam, was to station troops in blockhouses along the routes. At nighttime, when they could no longer communicate with one another (as they did with semaphores), each small garrison became vulnerable. What happened with blockhouses also pertained to towns. One guerrilla band was, in this way, able even to capture a small French-held city. To prevent the guerrillas from gaining local superiority, the French had to give up trying to control all Spanish territory and concentrate only on cities and supply routes. To keep from being overwhelmed at any given location, they had to employ large numbers of troops. One relatively small guerrilla *partida,* under the leadership of Javier Mina, is said to have occupied ten thousand French troops. This was more or less comparable to what Tito was able to do against troops of the German army and Ho Chi Minh did against the French. Finally, to keep the supply route from the French frontier to Madrid open, about a quarter of the entire French force had to be spread out along the way.

Long before the fall of Madrid, the French found that fighting guerrilla bands was like trying to push aside a puddle of water: when the French hand withdrew, they simply flowed back into place. As the fighting spread, the French realized that they could not garrison every village, so they experimented with various forms of what we call counterinsurgency. They offered amnesties to those who would surrender and, like the Americans later in Afghanistan, large payments for information. When these tactics did not work sufficiently, they tried various other means of increasing severity. Village officials were required to report who among their inhabitants had disappeared (and presumably joined the guerrillas); those who failed to report were heavily fined or, if thought to be complicit, were shot.

Where they could, the French enrolled Spaniards to fight Spaniards. Little is known about the *cuerpos francos* who played a role similar to the Loyalists in the American insurgency, but presumably they were driven by a combination of local animosities and French favor. Often the very excesses of the guerrillas created the motivation for such groups. Some of the guerrillas, to be sure, were mere outlaws, and at times all of them must have treated the villagers on whom they depended for food and shelter harshly. But most Spaniards thought of the anti-guerrillas as turncoats, *chacales* (jackals). Some of these turncoats were enrolled by the French into a counterforce, known as the Miquelet, to chase insurgents and punish the villagers who supplied them. The largest of these groups was about eight hundred men under Andrés Eguaguirre, a true brigand who did the French more harm than good. His force, known as the Volunteer Mobile Musketeers of Navarre, was to be the first of many failures of counterinsurgency. I will later discuss similar groups in Algeria, where the French created an Algerian force known as the *harkis* (native light troops) to carry out search and destroy operations or "regroup" the inhabitants.

The French regarded native opponents as invaders almost always have. They were mere "bandits." Centuries earlier, Pompey and Caesar had so regarded the early Spaniards, not as soldiers but as *latrocinio*. Later, British prime minister David Lloyd George would call the Irish Republican Army (IRA) "a small body of assassins, a real murder gang . . ." Prime Minister Winston Churchill would similarly term the Greek anti-Nazi partisans "miserable Greek banditti" and "treacherous, filthy beasts." Americans picked up the term for Filipinos.

Regarding one another as "illegal enemy combatants," the French and guerrillas fought one another ferociously. Prisoners were usually shot or hanged, as with the Germans and Partisans in Yugoslavia. One instance illustrates what was common: when three men in Pamplona were caught making cartridges, the French immediately hanged them. That night, a group of Spaniards apprehended three French soldiers, cut down the dead Spaniards, and hanged the Frenchmen in their places. The Spaniards attached a note on the chest of one Frenchman saying, "you take ours; we take yours." Locally and from time to time, efforts were made to contain the violence. Restraint rarely worked.

It wasn't only the French who came to fear the guerrillas. What remained of the former royal government feared that the call to arms—the "Manifesto of the Spanish Nation" issued by the popular *junta* government (the Supreme Central Assembly) and justifying the mobilization of civilians—would produce a truly revolutionary movement. Even if it was patriotic to fight the French, the royalists felt the guerrillas were engaged in illegal activities and, worse, were breaking the code that had controlled Spanish society for centuries: warfare was for aristocrats, not for peasants.

Consequently, as soon as the French began to withdraw, the Spanish rulers began to dismantle the guerrilla organizations. It proved far easier than anyone could have anticipated. With the French in retreat, there were no longer convoys to be looted. With French soldiers no longer threatening them, villagers who had supported the guerrilla bands no longer saw any need to do so. Contributions ceased to be given freely, and when demanded focused hostility on the guerrillas. As the royal government co-opted or purged the popular *juntas,* it reestablished its monopoly of taxation and customs so that the guerrilla leaders were no longer able to pay their followers. In desperation, the most successful of the guerrilla leaders, Javier Mina, appealed to the newly restored monarch, Fernando, to incorporate his disintegrating force into the Spanish army. In reply, Fernando issued a proclamation effectively proscribing the entire resistance: all organizations, both military and civil, that had not existed before the French invasion were to cease operation immediately. They had, in effect, been turned, as the French always insisted they were, into outlaws.

In conclusion, I turn to a contemporary treatise on warfare that was deeply influenced by Spain (and was a textbook for generations of cadets at the U.S. Military Academy at West Point). General Antoine Henri Jomini drew insights from the Spanish part of what must rank as the most wide-ranging military career of all time. What he saw in Spain he captures in two memorable phrases that go to the heart of guerrilla warfare: that they are often "wars of opinion" and nearly always "national wars."

Wars of opinion, Jomini wrote, come about when one state seeks to propagate its own doctrines or to crush the dogma of another state. He would have placed the current American neoconservative plan forcibly to impose "democracy" on other nations as a war of opinion. It would have horrified him. Such wars, he said, "enlist the worst passions, and become vindictive, cruel and terrible." They are fearful, he continued, "since the invading force not only is met by the armies of the enemy, but is exposed to the attacks of an exasperated people . . . History . . . appears to clearly demonstrate the danger of attacking an intensely-excited nation." Thus, attack and reprisal without restraint are inevitable.

Perhaps as bad but certainly more problematic are what he called national wars, since the invader has only his army whereas "his adversaries have an army and a people wholly or almost wholly in arms [so that] even the non-combatants have an interest in his ruin and accelerate it by every means in their power." Jomini was primarily concerned with the fate and actions of the major power, but what he was identifying were the roles of what Mao Tse-tung would later call "the sea" and "the fish." The sea is made up of the noncombatants who sustain the actual fighters, the fish. The fish are always few—the best number I have been able to find for Spain at any given time is about 13,000, although at times the number may have been double or triple this. This compares with perhaps 5,000–10,000 in Ireland, 11,000–13,000 in Algeria, 15,000–18,500 in Greece, and

15,000–20,000 in Iraq. "The sea" often numbered in the hundreds of thousands or the millions "who are determined to sustain their independence; then every step is disputed . . . each individual of whom conspires against the common enemy . . . No army, however disciplined, can contend successfully against such a system applied to a great nation . . . priests, women, and children throughout Spain plotted the murder of isolated soldiers."

And, as the French found in Spain and later in Algeria, the more effective, the more brutal they are in suppressing the general population, the more recruits the fighters gain, because in every instance in which a single combatant or even an innocent bystander is arrested, detained, wounded, or killed, a dozen of his relatives and friends are outraged. In consequence, the sea not only sustains the fish, providing them with food, information, and hiding places, but is their recruiting ground.

Thus, as Jomini suggests, commanders of regular armies are ill-advised to engage in such wars because they will lose them. Even where they appear to win, their successes are often only temporary. For an example of much publicized short-term success, I turn next to the Philippines.

CHAPTER 3

THE PHILIPPINE "INSURRECTION"

THE PHILIPPINE "INSURRECTION" OF 1900 WAS AMERICA'S FIRST TASTE of guerrilla war—other than campaigns against the Indians—since its own War of National Independence. The war actually began as a Philippine uprising against the Spanish colonial government shortly before the American invasion and was to fall into three phases—the "insurrection," the post–World War II civil war (also known as the Hukbalahap war), and the current Muslim attempt to break away to form an independent state. Here I will concentrate on the first two phases. The struggle in the Philippines illuminates two further characteristics of insurgency: the contrast between the short term and the long term in evaluating victory, and the efficacy of the techniques of counterinsurgency. The latter is of particular importance to Americans because the Philippine experience was sold to them as the way to win the war in Vietnam.

In considering the Philippines, one must begin with its complex physical and social geography: the country consists of some 7,107 islands, of which the two largest are Luzon in the north and Mindanao in the south. As would be expected from this geographical diversity, the roughly seven million people who inhabited the islands during the Spanish epoch were divided culturally, economically, and religiously. At least seventy languages were spoken; the ways of earning a living varied from fishing through forest hunting to intensive rice cultivation; and religiously, the population was mainly Muslim and Buddhist in the cities, with outlying areas primarily pagan. The southern population was converted to Islam in the fourteenth and fifteenth centuries by traders and teachers coming from Southeast Asia and the Indian Ocean. Islam spread quickly, and two Muslim kingdoms dominated the Philippines when the Spaniards first began to arrive in the sixteenth century. From their own historical experience, the Spaniards saw the native Muslims as just another species of the Arabs and Berbers, whom they had recently evicted from Spain and were then fighting in Africa; so they fastened the name Moro (Moor) upon the Muslim Filipinos. Spanish policy toward the Moros was simple: it was genocide. After three centuries of sporadic but intensely savage warfare, Muslims had been reduced from more than 50 percent to about 5 percent of the total population.

From another part of their experience, the Spaniards derived the model of colonial government and economic organization they imposed upon the Philippines. That was the system they developed in Nueva España (roughly modern Mexico), Nueva Andalucia (what later became New Mexico), and La Florida (what later became Louisiana, Florida, Alabama, Mississippi, and eastern Texas). Although the Philippines was administered from Nueva España, Spaniards moved throughout the islands. Particularly in Luzon, secular Spaniard soldiers and adventurers and some Church orders created vast landed estates, *encomiendas* or *haciendas,* in which they used the natives as agricultural workers to exploit the islands' vast natural resources. Meanwhile, friars and Church officials, backed by the colony's Spanish military force, set out to convert the pagan Malay people they called "Indios."

Like other colonial regimes, the Spaniards fostered the growth of a small cadre of natives, trained in Spanish, to man the lower rungs of their administration. Employing these partially Hispanicized natives was cheaper than importing Europeans; they were useful also because they could speak local languages. Over time, as they were influenced by Spanish culture, they aspired to more than menial roles in the administration and second-class status in the society. After the middle of the nineteenth century, some Filipinos made their way to Europe, where they studied in Spanish and German universities. When they returned, these men, known in Filipino history as the *ilustrados*, formed a small, elite group that found the discrimination the Spaniards imposed upon them particularly galling because they had been treated well in Europe. Briefly encouraged by a relatively liberal Spanish governor, they began to assert their aspirations to be treated like Spaniards. In 1871, however, that governor was replaced by a martinet who reimposed a rigorous system of apartheid and repression. The sequence of discrimination to liberalism to discrimination provoked expressions of resentment, and the new governor feared their anger might incite unrest. So he had those *ilustrados* most identified with the reforms exiled, leaving behind only a memory of the reform period.

It was this memory on which a leader of the next generation, José Rizal, a young Filipino of partly Chinese ancestry, drew. Like some of the figures of the previous generation, Rizal had studied in Europe, where he had learned to expect respect for his professional skills. Returning to the Philippines in 1892, he was immediately reminded of the realities of Spanish rule. His reaction, mild as it was, was to found a small literary society known as La Liga Filipina. La Liga did not advocate violence, but it did discuss the desire of its members for more equal relations with Spain and restraints on the power of the religious orders. While social programs were not high on their priorities, the Liga members also debated the effects that could come from abolition of the system of forced labor (known in the Philippines as *polo*) and the system under which peasants' products were requisitioned (known as *vandala*) by the government.

Although La Liga had few members and they were hardly revolutionary, the government regarded them as a mortal threat and moved to destroy them. Rizal and all the members of the group whom the Spaniards could identify were imprisoned or exiled. One who got away was Andrés Bonifacio. Bonifacio was the odd man out. He had little education of any sort and had never been abroad. Moreover, he was of much more humble social origin than Rizal or the other members of La Liga; but he was a much more determined, practical, and radical man than any of Rizal's circle. Rizal's removal gave him an opening. Putting aside talk of social acceptance or even incorporation into the Spanish government, he determined to gain Philippine independence. He believed that the Spanish would never pay any attention to literary societies or the debates of men they treated as mere clerks. The only action the Spaniards would understand or credit, he decided, was armed force. Since little was left of La Liga Filipina after the governor's suppression, Bonifacio set about creating a new and very different sort of organization. The change was symbolized by the fact that its name was not Spanish but Tagalog: Katipunan.

The Katipunan (usually translated as "Patriots' League") was a secret organization—it had to be if its leaders were to stay out of prison. For safety, it was organized on a classical revolutionary "cell" pattern. Each member of Bonifacio's original group was ordered to select two followers whose identity would not be known to each other. Each of them in turn would select two more and so on. For nearly four years, the existence of the Katipunan was unknown to the Spaniards; then, a priest, who had heard a confession from the sister of one of the leaders, denounced the organization to the Spanish authorities. Unable to catch Bonifacio, they executed Rizal although he had nothing to do with Katipunan; his execution gave Bonifacio's movement what it had lacked, a martyr.

Imagining a vast plot extending to the Philippines from Cuba, where rebellion had broken out in February 1895, a newly appointed military governor set about destroying Katipunan. If the numbers of people arrested, tortured, and executed—about four thousand—can be taken as a guide, Bonifacio had done a remarkable job of creating his secret organization in just four years. Before they went down in defeat, Katipunan fought determinedly against both the government and the Church. Although most members of the society and their peasant supporters were at least nominally Catholic, they evinced great hatred for the Church. Probably viewing it not only as a "parallel" government but also as an exploiting landlord and as an instrument designed to take away their culture—much as the Pueblo people had viewed the Church in New Mexico—they targeted Church installations and its "foot soldiers," the ubiquitous friars, for savage reprisals. The carnage was spectacular. The peasants enrolled by Katipunan fought valiantly, but they were virtually unarmed and untrained. Moreover, they made the mistake of trying to fight the Spanish army in a conventional way. Consequently, they were defeated in every encounter.

Pounded by the Spaniards and driven apart to remote hiding places, the Katipunan society and the underground government it had organized split into various mutually hostile factions. One of these, led by another part-Chinese Filipino, Emilio Aguinaldo, captured and killed Bonifacio. Whether their hostility was personal, political, or paid for has never been determined. Aguinaldo then reached a deal with the Spanish governor: the insurgency would stop, the peasant soldiers would go home, and he would be allowed to go into exile with a purse full of gold. It was not so much that Spain had won this first uprising as that the rebels, by fighting with one another, had lost. The suicide of this insurgency was not unique; as we shall see, it was to be repeated halfway around the world and half a century later in Greece.

What proved disastrous to the Spanish position in the Philippines was not Katipunan but the coincidence of the revolt there with the outward thrust of America under presidents Grover Cleveland (during whose administration Hawaii was taken over in an American-organized coup d'état) and William McKinley (whose administration brought America into conflict with Spain). The first move under McKinley was against the Spaniards in Cuba. The explosion of the USS *Maine* on February 15, 1898, now believed to have been an accident but then blamed on Spain, provided a popular cause; using it wildly, irresponsibly, but profitably, the press whipped up a call for war. One of the few who urged restraint, Republican Senator John Coit Spooner of Wisconsin, sounded a very modern note when he commented that "perhaps the President could have worked out the business without war, but the current was too strong, the demagogues too numerous, the fall elections too near." Congress voted for war in April.

During the time the crisis leading to the war with Spain was building, Aguinaldo was in gilded exile in Hong Kong. There he met with several American officials and offered to assist an invasion of the Philippines. Allegedly, in return he received assurances that the United States, being "a great and rich nation . . . neither needs nor desires colonies . . ." The ranking American officer with whom he met, Commodore George Dewey, is reported to have told him, "I have no authority [to commit the American government]; but there is no doubt if you cooperate with us and assist us by fighting the common enemy, that you will be granted your freedom the same as the Cubans will be." That was assurance enough for Aguinaldo. He then acted on what he thought was his part of the bargain. On an American naval vessel, he returned to the Philippines.

Taking advantage of the American naval defeat of the Spanish fleet at Manila Bay on May 1, 1898, and the subsequent withdrawal of the Spanish army to Manila, Aguinaldo gathered together a group of insurgents large enough to take over the whole of Luzon island except for Manila and proclaimed an independent republic. What he thought he had been assured in Hong Kong was confirmed in Washington, where President McKinley publicly repudiated "forcible annexation . . . [as] criminal aggression." When informed of McKinley's statement, Aguinaldo proclaimed the arriving American troops to be the Filipinos' "redeemers" and welcomed them with open arms.

The Spanish army was quickly defeated and evacuated; the Spanish bureaucracy collapsed, and the Church was in temporary eclipse. Although Aguinaldo's movement tried to supply some degree of law and order, it was ill-equipped to do so. Having been excluded from the Spanish administration, Filipinos had little experience. And, largely excluded from land ownership and commerce, they had little money. So it was inevitable that when the American "Philippines Commission" reported to President McKinley, it would argue that unless Americans actually took over the country it would fall into chaos. In attempts to counter this argument, Aguinaldo sought to organize—or at least proclaim—ventures that would show that his people could actually rule themselves. In gestures he thought would appeal to the Americans, he prepared an American-style constitution and put a newly proclaimed American-style congress into session.

These gestures were brushed aside by the local American officials. What they saw—the record seems to confirm also that what they wanted to see—convinced them that Aguinaldo's movement was just a façade and that the Filipinos were just like the American Indians, sometimes amusing, always murderous, only partially human creatures. At best, they were what Rudyard Kipling was then describing as "The White Man's Burden . . . Half-devil and half-child." They had no doubt that the Filipinos had to be ruled with a firm hand. And that could only be an American hand. Meanwhile in Washington, McKinley "went down on my knees and prayed Almighty God for light and guidance." God replied, he said, ". . . to take them all . . . and then I went to bed, and went to sleep, and slept soundly." American forces were committed to rule "Your new-caught, sullen peoples" and to fight God's fight. That would not be the last time an American president proclaimed that he was acting on God's order.

When the Filipinos learned that in the December 10, 1898, peace treaty with Spain, America had "bought" the Philippines for $20 million, the Filipinos stopped cooperating with the American troops. America had decided to join the imperialists' race to "color the world map."

To color its part of the world map, the expeditionary force's objective was changed from defeat of Spain to conquest and occupation of the Philippines. As generals nearly always do, its commander warned that it was inadequate for the task. Influenced by the textbook he had studied at West Point, Jomini's *The Art of War,* he predicted that the majority of the Filipinos "will regard us with the intense hatred both of race and religion." He was right on both scores. The troops *were* inadequate, and even before the first major engagements, Filipinos had learned to hate Americans.

Violent resistance from virtually the whole population was triggered by an American killing a Filipino soldier on February 4, 1899. Fighting quickly became as bitter as the wars of extermination against the native Americans. Having turned down Aguinaldo's offer of negotiation, the Americans launched an attack on his ragtag army in which they killed several thousand Filipinos.

While they had studied Jomini's *The Art of War,* it is not known whether the American commanders had read the work of the contemporary English theorist of colonial warfare, Captain (later Major General Sir) Charles E. Callwell. The first edition of his book *Small Wars* had been published in London in 1896 and a second edition in 1899. Whether they had read him or not, they followed the advice he laid out. It was to "chastise the rebels in their homes . . . to bring them to reason he must reach them through their crops, their flocks, and their property . . . depriving them of their belongings or burning their dwellings." Only thus could the insurgents be forced into battles where they could be destroyed by the superior forces of the occupier. Sending out "flying columns" as Callwell had advised, the Americans destroyed dozens of villages. They made no attempt to "win the hearts and minds" of the natives. That notion was alien to Callwell and other practitioners of colonial warfare. Wherever the incoming Americans met Filipinos, including those who offered to help them, they customarily referred to them, even in their presence, with the racial taunts they had brought from America. To them the Filipinos were "niggers" or other derogatory names—geezers, coons, gooks, wops, ragheads—Americans adopt so easily. In the Philippines, the natives were "gugus."

Furious and disheartened, but still far from united as a nation, at least those Filipinos influenced by Aguinaldo were forced to learn guerrilla warfare. Aguinaldo seems to have had an instinctive understanding of informal combat. More important still, he had an understanding of the politics of insurgency. His minuscule "army" fractured into bands that melted into the general population until opportunities for ambush or small-scale attack appeared.

Opportunities for attack appeared quickly. Consequently, as in the later Vietnam War, the limited first American step into the "quagmire" required a second more massive step. To get the military force it belatedly realized it would need to support its first contingent, the American government (as it would later do in Iraq) called up large numbers of National Guard troopers. For these soldiers and for nearly all their officers, the Philippines campaign was the first venture outside North America. And the world they found was to them almost totally alien. They hated it. Thousands immediately contracted tropical diseases against which they had no immunities and against which medical science then had few preventives or cures. They found the climate almost unbearable and the food completely uneatable. Worse, they could not understand why the Filipinos disliked them. In short, they were ill-prepared in every sense for what they faced—as a contemporary humorist scoffed, prior to the invasion, they had not known "whether 'the Philippines' were islands or canned goods . . ."

Atrocities soon became the order of the day for both sides. U.S. senators were told of the tortures commonly employed by American troops—the favorite, sounding a very modern note, was the "water cure." In it buckets of water were forced down a captive's throat and then forced out by men kneeling on his stomach. When they could retaliate, Filipino insurgents burned or hacked to death captured Americans. Terror bred terror.

Also striking a modern note, where reporters are either "embedded" or otherwise have their access to events restricted, the senior U.S. commanders tried hard to prevent news of atrocities and casualties from reaching the American public. The commanding American general, Ewell Otis, personally altered the dispatches sent out by journalists. His chief censor told the outraged reporters, "My instructions are to shut off everything that could hurt McKinley's administration," and when the reporters confronted Otis, "Otis did not deny suppressing the facts. He had to do it, he insisted to shield the people from distortions and sensationalism." So what they had themselves witnessed, "such as American soldiers bayoneting wounded *amigos* [Filipinos], the looting of homes and churches, and so on," was not to be reported.

By the year 1900, America had 150,000 soldiers in the Philippines (almost exactly the number America had in Iraq in 2007). As many of his successors would do in Vietnam and Iraq, the American commander announced that the war was already over: there "will be no more real fighting . . . [just] little skirmishes which amount to nothing . . ." "Two years and six thousand U.S. casualties later . . ." the general's proclamation of victory would come at least temporarily true.

In the meantime, to try to square what was actually happening with what the administration wanted the American public to believe, the high command tried to keep American casualties low. So it began to recruit a native army, as the French had done in Spain and were later to do in Algeria. In the Philippines, these special forces were known as the "Scouts." Divided into small flying columns, as Callwell had advocated and as the British later did with their "shock companies" in Ireland and South Africa, these groups raided villages, captured men they could turn by bribery or torture, and carried out savage reprisals. In a particularly daring raid, one small detachment penetrated Aguinaldo's hideout, took him prisoner, and convinced him—he had previously shown himself particularly susceptible to such convincing in return for money—to declare the surrender of his partisans.

What happened then was what almost half a century later ensured the collapse of the Greek EAM/ELAS insurgency: the insurgents split into two camps. Those who followed Aguinaldo put down their arms, while the others, deprived of Aguinaldo's leadership, lost coordination and were driven into places where they could not supply themselves. The Filipino population was confused. Since Aguinaldo was the best-known leader, most followed his example, relieved that their struggle had ended. Thus, as Mao Tse-tung would have put it, the "sea" dried around the "fish": villagers stopped supplying and hiding the combatants. Loss of popular support was mortal to the resistance. The process was speeded up and spread by various forms of counterinsurgency: those Filipinos who were thought likely to continue to support the insurgents were "regrouped" to places where they were cut off from the still-active insurgents and watched helplessly as their villages were burned and their food supplies confiscated; to prevent the insurgents from getting arms, even the newly created Filipino police were disarmed. The suppression was as brutal as in any guerrilla war.

The American governor, the later Republican president William Howard Taft, added a third ingredient to the campaign against the guerrillas: counterinsurgency. Unlike the French in Spain whose counterinsurgency consisted of little more than terror, Taft decided to win over the rebels. He would make it worthwhile for them to "turn." He shrewdly instituted a program aimed to ameliorate what was already recognized as the most painful legacy of Spanish and Church rule, the alienation of the vast majority of Filipinos from the land. Taft bought from the Church some 640 square miles of valuable agricultural lands and began to distribute them on easy terms and in small plots to farmers who showed their willingness to cooperate with the American administration. He also began what would later be termed a "civic action program" in education and public health by bringing to the Philippines more than a thousand Americans in an early model of the Kennedy-era Peace Corps. Perhaps even more intelligently, he reduced and withdrew to less conspicuous places the large American military presence, so that "Filipinos should suppress Filipino disturbances and arrest Filipino outlaws." In the new order, as in other counterinsurgencies, those who resisted as well as those who aided them were treated not as patriots but as bandits (*ladrones*) to be hanged or incarcerated for life.

The cost of the invasion and occupation had been high: an estimated two hundred thousand Filipino civilians and about twenty thousand insurgents had been killed, while four thousand Americans had died in combat and many more would die of the diseases they had contracted during the war.

Contemporaries believed that the insurrection was over. Events were to prove them wrong. Despite Taft's "civic action program" and the land distribution scheme, the issues that had impelled the insurgency had not been satisfactorily addressed. The Spanish flag had been hauled down, but an American flag had replaced it. It would not be lowered for nearly half a century. And during that half century, the Philippine population would become increasingly educated and "politicized" so that colonial status would become more rather than less galling. Hopes had been raised by Taft's gestures, but the hunger for land had not been satisfied. Most Filipinos were still wretchedly poor. The disparity between the rich and the poor grew. In the eyes of large numbers of Filipinos, the exploitative landlords and the American-controlled national government were just two arms of the same body. Both used the newly created national police, the Philippine Constabulary (PC), as a truncheon to beat down peasant resistance. As one historian put it, the constabulary "was practically an army for the landed elites. When landlords needed help, whether to guard their fields against tenants who wanted to harvest early or to barge in on peasants' meetings, they had only to ask the local PC commander to send detachments." Thus, although too weak and still too disorganized to resist on a large scale, Filipinos continued small-scale and sporadic outbreaks throughout the years before the Second World War.

During the Second World War, a new cause, expulsion of the Japanese, was picked up by a new generation of Filipinos. Because the Philippine Constabulary and the civil administration as well as the large landowners (the *hacienderos*) generally collaborated with the Japanese, the war in the underground fostered new notions of political organization and new leaders outside the governing circles. And as angry men ambushed patrols of Japanese soldiers and Filipino police to get arms, the identification of the old regime with the Japanese invaders became fixed: jointly they were the enemy.

As in occupied Europe, resistance to this Filipino-Japanese enemy was begun haphazardly and on a small scale by scattered groups whose members had been harmed by the invader. Then, as in Yugoslavia and Greece, the small groups began to coalesce into a movement. The main group in the resistance, the Hukbo ng Bayan laban sa Hapon (the People's Anti-Japanese Army, known as the Hukbalahap or Huks), grew out of a prewar agrarian movement known as the Kalipunang Pambansa ng mga Magsasaka sa Pilipinas (National Society of Peasants in the Philippines, or KPMP), which had struggled to end abuse of the peasants. So the Hukbalahap was already recognized as a social movement among at least the poorer Filipinos before the coming of the Japanese. It would be legitimized politically by opposition to the Japanese, some five thousand of whose soldiers it killed during the occupation. Encouragement for it came both from a nationalist sentiment and from anger at Japanese atrocities. The tens of thousands of KPMP members, their relatives, and friends became, metaphorically, the sea in which the Huk combatants swam.

Since the Huks almost wholly depended on the noncombatants, they guarded them carefully. Usually guerrilla units operated close to their homes and so moved among relatives, neighbors, and friends. Any guerrilla who harmed or insulted a peasant farmer was severely punished. Whenever possible, also, the Huks paid for the supplies given them by the general population. Consequently, as one informant put it, "Almost everybody in [the province of] San Ricardo was part of the Hukbalahap. If not actually carrying a gun, then they gave information and food to the Huk soldiers."

Opposed to dealing with the Hukbalahap, which American counterintelligence regarded as a Communist front, but also anxious to undermine the Japanese occupation and prepare for the American return, the U.S. Army set up a rival organization known as U.S. Armed Forces of the Far East Squadron 311. In the eyes of the Hukbalahap and their supporters, this group was compromised from the very beginning because it formed an association with and relied upon Filipino officials and large landowners who collaborated with the Japanese. Unlike the Hukbalahap, Squadron 311 had no political organization and seems to have been comparatively ineffectual. Then, in the last days of the occupation, groups of collaborators and members of the Philippine Constabulary began to desert the Japanese. Many found a haven with Squadron 311, where they affirmed their allegiance to America. Soon these groups were at daggers drawn with the Hukbalahap, who regarded them as traitors. In this, as we shall see, their activities and organization resembled the British-supported reconstituted gendarmerie in postwar Greece.

The Hukbalahap emerged from the war with a new sense of purpose and power. Its numbers had grown from fewer than three hundred in the spring of 1942 to between ten thousand and twelve thousand in the fall of 1944. They had fought the Japanese, as the official history of the American army put it, "furiously," rescued American aviators, and controlled significant sections of the islands; consequently, they were convinced that they were, and should be regarded as, America's allies. Like their fathers, these Filipinos believed that they could work with the Americans to express their desire for independence and social reform. And like their fathers, they were quickly disappointed. Indeed, there is a parallel on the American side of this echo of generations. The key American figure, General Douglas MacArthur, was the son of General Arthur MacArthur who had helped to shape the original American policy. And, more to the point, Douglas MacArthur had spent from 1935 to 1941, as "field marshal" of the army of the Philippines, working for and with the Filipino landowning and business elite whom the new insurgents wished to topple.

MacArthur's background showed in his first actions: he disarmed as many of the Huk guerrillas as could be trapped, arrested their leaders, and arranged that men of the old regime, even those known to have worked for the Japanese, be appointed to senior government posts. His policy seemed almost calculated to provoke the guerrillas, since more than a million people had been killed during the Japanese occupation. Undeterred, MacArthur pardoned Manuel Roxas, the best known and most senior Filipino collaborator, and threw American support (and money) into Roxas's campaign to become Philippine president in 1946. Then on July 4, 1946, he moved to undercut the Hukbalahap movement by what in most circumstances would have been a winning ploy, recognizing Philippine independence. But the Huks regarded it as just a clever move and did not believe it meant real independence.

It was almost inevitable that in response to MacArthur's policies, those members of the Hukbalahap movement who were still at large threw in their lot with the only remaining opposition group, the Partido Komunista ng Pilipinas (the Philippine Communist Party, or PKP). Meanwhile, the new Roxas Administration began its move against the opposition by refusing to allow into the legislature the six Huks who had been elected; at the same time, it began a crackdown, known as the "Pacification Program," on opposition groups throughout the country. Juan Feleo and other senior leaders of the Hukbalahap movement were assassinated, presumably on government orders, in the summer of 1946.

Feleo's assassination was the trigger that fired the new outbreak of insurgency, but the underlying causes were deep and by then more or less traditional. As Richard Kessler commented in *Rebellion and Repression in the Philippines,* "The Huks emerged as a revolutionary force against this postwar backdrop of extreme political animosity, a ready supply of weapons, and economic deprivation. Peasant earnings did not reach prewar levels [for many years] . . . Given the history of grievances, especially in central Luzon, and the conditions then prevailing, it would have been strange if rebellion had not broken out." It did. After Feleo's murder, the Hukalahap leader Luis Taruc stopped trying to work out a compromise with the government and took to the hills with the peasant insurgents.

Between 1946 and 1949, the Philippines was convulsed by civil war. During those years, about 25,000 government troops and paramilitary auxiliaries fought against less than half that many insurgents. The insurgents, however, were supported by at least 150,000 Filipinos. These were peasants, of whom most lived in six thousand square miles of rice fields in the center of Luzon island where they worked as sharecroppers for absentee owners.

Military suppression did not work. After four years of fighting, there were more insurgents and a much larger group of supporters. So what to do about it? There were two quite different options: try harder to suppress it or remove the causes. The choice was not really made on an analysis of the conflict in the Philippines—as it was later not made in Vietnam or Greece—but in the context of the evolving cold war. The Soviet hand was seen everywhere by American strategists and statesmen who knew nothing about the local, even recent, causes of conflict. So how was the choice shaped and to what effect?

Enter Colonel Edward Lansdale, who, on behalf of the CIA, aspired to be a modern Machiavelli and found his "prince" in the then minister of defense of the Philippines, Ramón Magsaysay. Unlike many of the men of his class, Magsaysay had actually played a small role in the anti-Japanese movement and claimed to be an anti-colonialist. But he did not fight his way into his role as a Philippine patriot; he was appointed by the Americans. Apparently, he then had no coherent political beliefs. More important, unlike Ho Chi Minh in contemporary Vietnam, Magsaysay had no political organization. It was organization and method of operation that Lansdale aspired to give him.

Lansdale succeeded in a way that seemed almost miraculous to Eisenhower's secretary of state, John Foster Dulles, and subsequently to President Kennedy. Kennedy saw Lansdale not so much as a Machiavelli but as an American "007"—James Bond. Most American readers would know him as "Colonel Hillandale," in *The Ugly American*. In fact, he was a precursor in the Philippine context of a political strategist like President Bush's Karl Rove. But he was a Karl Rove with an army, a police force, and a national budget backed by the United States, and he operated in a civil war that relieved him of all restraints of law. Building upon these assets and driven by an almost messianic zeal, Lansdale literally made Magsaysay. He even co-opted Magsaysay as his roommate in an American military compound so he could tutor him more effectively both day and night. He could and did claim that he had "saved" the Philippines that America wanted from its Communist enemies. So miraculous did his venture seem that it was natural, even inevitable, that as America began to replace France in Indochina, it would send its Machiavelli–James Bond–Colonel Hillandale–Karl Rove over to repeat his performance there.

In his account of his venture in the Philippines, *In the Midst of Wars,* realist that he was, Lansdale minutely reports on the corruption, tyranny, and deception of the Philippine government and contrasts it with the ways in which the Huks conducted themselves. He vividly reports on how the "farmers were losing everything they possessed through high rentals, decisions in the land courts that always favored the landlord, the usurious money lenders who charged farmers mounting interest on the original small debts of their fathers. The townspeople had gone to the polls in 1949 and voted for representatives who could change all this. What had happened? Nothing. They had been cheated. The election had been rigged, bringing the corrupt to power. Of course the people were joining and helping the Huks. The Huks were right."

So what did Lansdale favor? In the Philippine context, the Huks, since the Philippine government "was rotten with corruption." But, in the context of the cold war, that was impossible. Caught up in the logic of the cold war, he had concluded, as had the whole American government of the time, that the Hukbalahap was, even had to be, a Communist conspiracy. True, it was not "Communist" until American policy drove the Huks into Communist arms—but for him, such facts, while deplorable, had to take second place to the reality of the larger war, the cold war with the Soviet Union, in which the Philippines was a small but important theater.

So, in the very next page after his portrayal of the ugly reality of Philippine politics, Lansdale switches his moral perspective to discuss how to "upgrade" the military intelligence service so that it could be more effectively employed in supporting those same corrupt officials and oppressive landlords. Unconsciously or at least without attribution, Lansdale took Magsaysay back to the "civic action" policy of William Howard Taft. And to those who were ignorant of Philippine history, what he proposed seemed stunningly, brilliantly innovative. What the Taft-Lansdale-Magsaysay program amounted to was a dual set of actions. On the one hand, the military had to be cajoled or encouraged to fight. Regular forces had to be "stiffened" with what were called "scout-ranger" teams, small, flexible, and mobile. They were composed usually of just one officer and four soldiers. Where more soldiers were required, they were provided by what came to be called "battalion combat teams." These groups should seek out, ambush, and destroy Huk insurgents. Taking advantage of their mobility and firepower, they were highly successful. From 1950 to 1955 they were credited with killing 6,874 Huks and capturing an additional 4,702. Trapped or starved into submission, a further 9,458 were said to have surrendered.

Military action was the fist. But, borrowing from Taft, Lansdale realized that force alone was insufficient. In a land where nearly half the population was below the poverty line, at least the *hope* of improvement in living standards was necessary. The easiest way to inject money into the economy was foreign aid. America supplied most, but European countries also helped. And by favoring or at least allowing unemployed Filipinos to work abroad, billions of remittance dollars poured into families at home. An Economic Development Corps (EDCOR) began in 1951 to parcel out farms from public lands "as a start toward dimming the Huk rallying cry of 'land for the landless.'" Huks began to surrender on the promise of acquiring plots of land with tools and seeds on concessionary terms.

Under Lansdale's tutoring, Magsaysay became the star of Philippine political life. Constantly on the move, listening to the complaints of the common people, and undertaking in countless small ways to assuage, if not solve, their problems, he offered what Filipinos found as revolutionary as the siren call of the Communists. He was elected to the presidency in 1953.

With the combination of Magsaysay's and Lansdale's methods, the removal of the visible sign of American hegemony, massive foreign aid, the removal of a large part of the Philippine labor force for whom there were no jobs domestically, and their remittances from abroad to their families, the insurrection faded away. Or at least it did so in part and temporarily. After Magsaysay's death in a plane crash in 1957, Philippine political life returned to the more traditional patterns of corruption, great differences between the rich elite and the poor peasantry, and unresolved issues like the position of the remnants of the Muslim population in the far south. In addition to the Muslim movements in the southern islands (the Moro National Liberation Front, the Moro Islamic Liberation Front, and the much smaller Abu Sayyaf), a "dirty war" is in progress in the north, especially on Luzon island in what looked like a resurgence of a movement similar to the Hukbalahop, the New People's Army. As Philip Bowring observed, "The police and military are often seen as being in league with landlords against tenants and workers . . . the wider social malaise that the dirty war represents will remain." All that was in the future. But what mattered then in the larger picture was that the Philippines—both the insurgency and the counterinsurgency—was taken as a model for the events already unfolding in Vietnam.

CHAPTER 4

THE IRISH STRUGGLE FOR INDEPENDENCE

FOR NEARLY A THOUSAND YEARS, THE IRISH HAVE STRUGGLED AGAINST, been overwhelmed by, or been forced into accommodating foreign rule. Consequently, their history is a virtual catalog of the various aspects of resistance and insurgency. As I will show, the Irish experience has involved invasion and colonization, guerrilla warfare and counterinsurgency, genocide and emigration, partition and "nation building," terrorism and internment. Periods of relative calm have been interspersed by savage conflict. As the 1956 *Handbook for Volunteers of the Irish Republican Army* asserts: "No nation has a greater tradition of guerrilla warfare than Ireland." The costs in shattered lives have been tragic, and the monetary costs have been almost ruinous to generation after generation. So there is much that can be learned from the rich lode of Irish history.

The war of the Irish for independence can claim to be the longest lasting in world history. We may date it to roughly a century after the 1066 Norman invasion of England when some of the descendants of the Normans, unsatisfied by the spoils of their invasion of England, moved across to Ireland. In the twelfth century, the Anglo-Norman king Henry II, with the blessings of the papacy, took the title "Lord of Ireland." His knights overwhelmed the primitive Irish and seized much of southern Ireland, but they were too few to colonize what they conquered. Gradually the Irish fought back and regained territory so that what the English called "the obedient shires" became just a small coastal strip north of Dublin. The English government then lost interest in Ireland, and the descendants of the conquerors took Irish wives and begat increasingly Irish children. By the fourteenth century the "Old English" had faded into the Irish population. Thus Ireland languished, at least strategically, until the time of Henry VIII.

Henry VIII was driven into Irish affairs by his fear of the superpower of the time, Spain. When one of the fifty or so Irish "kings" appeared likely to make Ireland into a Spanish outpost, Henry began a move to extend the "Englysshe Pale" (the already subdued area centered on Dublin) over the whole of "Great Irishry." Henry's main weapon was control over land: he forced the Irish chieftains to surrender their lands to the English Crown, which then "re-granted" them to the chiefs conditionally. Only so long as they were loyal to the English Crown could they keep the lands. It was a shrewd move, but it did not solve England's Irish problem.

Observing that problem, the Spaniards saw an opportunity to turn Ireland into England's unwinnable war, its quagmire. Spain was moved to adopt this policy because English pirates were preying on its economic lifeline, its transatlantic treasure fleets that brought back gold and silver from Mexico and Peru. After futile and expensive attempts to halt piracy in the Caribbean and along the North American coast, the Spanish government concluded that it had to attack the pirates in their "nest," England. That was the reason for the Great Armada of 1588. When that attack failed, the Spanish government decided on a diversion to Ireland. Even if the Irish could not defeat the English, the Spanish government judged, they could distract and weaken England. So Spain sent yet another fleet to aid the Irish in 1597 and a third in 1601. Like the first armada of 1588, they failed in fire and storm, but the survivors who made it ashore served as advisers, to use the modern subterfuge, to teach the Irish insurgents the use of cannon. Then, somewhat more equal, both sides fought with a ferocity seldom matched until the twentieth century.

To stanch the drain on the treasury and to prevent Ireland from becoming a Spanish base, first Queen Mary and then Queen Elizabeth undertook to change the very nature of Ireland. They employed two techniques that echo down the ages and still shape Irish affairs. The first was military. They hunted down and killed the Spanish soldiers who had been sent to train and fight alongside the Irish. Irish resistance justified a campaign of brutal search and destroy operations. These operations were harbingers of campaigns the British army and colonial auxiliaries later mounted in the New World against native Americans, razing houses, burning fields of grain, killing livestock, spoiling caches of food, and enslaving or slaughtering men, women, and children. Irish prisoners of war were often shot, hanged, or drowned, while their women and children were used for target practice. In one still bitterly remembered incident, the Earl of Essex lured the Irish chiefs and their wives to a peace conference at Belfast castle; there, as the Irish national epic recounts, after three days of banquet, Essex had them seized, immediately killed most of them, and sent their leader and his wife to Dublin, where they were "cut in quarters. Such was the end of their feast." Despite English ferocity, or perhaps because of it, the Irish fought on. Knowing the fate that awaited them upon surrender, they thought they had no choice. The English reaction was genocide: "We spare none of what quality or sex soever . . ." Grimly stated to justify the slaughter of children, the English worked on the principle, as one of their generals put it, that "nits make lice." The available technology did not make genocide possible so the English fell back on "removal." Inexorably they drove the Irish survivors out of their agricultural lands into the wilds of the north and west, where many starved and the survivors were reduced to dire poverty. But, no matter how brutal, the military campaigns proved ineffective. The English were not able to achieve "security." It was at that point the English hit upon the second of their two policies.

Under Elizabeth's cousin and successor, King James, the British exchanged *imperialism* for *colonialism:* that is, they moved from merely *conquering* Ireland to *replacing* the Irish. In 1607, the same year as the first successful move to Virginia, the British sent commissions to Ireland to survey the country and organize it into "plantations" for incoming English and Scots colonists. The "plantation" of these people was the origin of the most acute problem of today's Ireland, bitterly divided communities in the six counties of the north, Ulster.

Ulster became the base where the English (and their Scots protégés) were safe and from which British armies could sally forth to meet any challenge. But the English knew well how expensive armies are, so they implemented two subsidiary policies to make military operations unnecessary. First, as they would later do to the native American Indians in the New World, they began to remove the Irish to reservations, and, second, they began a program of what today we would call sanctions. Implementing a series of economic restrictions known as the Navigation Acts, they sought from the 1660s to weaken Irish agriculture and cripple Irish industry. The policy also had cultural dimensions. Religiously, the grip of the Church of England was tightened on the Catholic population, and efforts were made to force the Irish to speak only English. The aim was to break Ireland's cultural tradition. Only in the Catholic institutions was a memory, pale though it was, kept of the Irish past. Contemporary morale plummeted. As a result, as Jonathan Swift observed, the Irish were "dying and rotting by cold and famine and filth and vermin." Thus began the process that would devastate Ireland in the eighteenth century.

As the British wore the Irish down, they fastened upon the country a regime, as the English historian Roy Porter has written, so that Ireland was "bullied and bled by a frequently absentee Protestant landlord class," which was awarded lavish bonuses (paid for by the Irish) for service to the British Crown. The Irish parliament was emasculated in 1719 by a "Declaratory Act" on which the 1766 Declaratory Act dealing with the American colonies would be patterned. Under it, the Irish parliament was denationalized: among its 307 members in 1775, only 44 Irish names appeared; the rest were either English or descendants of the Scots-Irish settlers. Only 72 of the members were elected; the rest were appointees of the English viceroy. Roman Catholics, who made up more than three-quarters of the population, were totally disenfranchised, and at least half of the Protestant Irish were also excluded by the 1704 Anglican sacramental "Test Act," which imposed the Anglican creed as a prerequisite for the franchise. In the course of the nineteenth century, the Irish mostly forgot their native language, but hung on to their religion.

In the terrible famines between 1845 and 1855, when the potato crop—the basic food for most of the population—failed, nearly a million people died and another two million immigrated to America. Those who remained lived in conditions unmatched for misery in Europe except among Russian serfs. The British government took little note of the widespread suffering. The only charity—which the affluent then regarded as corruption of the poor—was provided by workhouses; in them conditions were purposefully harsh, but grim as they were, they did provide inmates with some food. However, they could accommodate only about one Irishman in every fifty or sixty. For the rest, there was just hunger. Conditions were made worse by the larger, mainly Protestant, landlords' eviction of their almost entirely Catholic tenant farmers. Only the importation of corn (maize), provided from America by Irish émigrés, prevented total destruction of the population. Gradually and reluctantly, the British government undertook public works projects to pump some money into the economy, but its effort was pitifully inadequate.

By the beginning of the twentieth century, Ireland was not so much a nation as a nostalgia. While small groups of intellectuals met to discuss their heritage—and a few bold or desperate men even carried out acts of terrorism against those they identified as their oppressors—few Irishmen were interested. Their passivity arose, I think, from two causes. In the first place, neither in Ireland nor elsewhere do broken people revolt. Revolt requires a modicum of resources and the energy given by hope. By the middle of the nineteenth century, few Irishmen had either resources or hope. The second cause was the lack of a common cause sufficient to form the basis for organization. Religion proved inadequate because "the potato blight also exposed and exacerbated the social divisions *within* Catholic Ireland, and many of the hardships poor Catholics endured came at the hands of their more fortunate coreligionists. Catholic landlords were no more indulgent toward tenants than were Protestant proprietors; indeed, to a *Catholic* landlord in Tipperary belongs the credit for inventing a machine for the cheaper and more expeditious unroofing and demolishing of tenants' homes." Ironically, Irish nationalism was the gift of the Protestant community of Ulster.

It was the Protestant community that took the first step toward a form of political association when nearly half a million people signed the 1912 Ulster Covenant. Those who signed it aimed to prevent Britain from spinning off Ireland by implementing the "Home Rule" bill that Parliament had just passed. They sought to differentiate themselves politically as well as culturally from the Irish, a colonial people whom they despised, and to maintain their status as Englishmen. To thwart the new law, one hundred thousand Protestants formed themselves into an armed national militia, the Ulster Volunteer Force.

Formation of the Ulster Volunteer Force was a turning point in Irish history. As the Irish journalist Tim Pat Coogan has written, "Had the Protestants of the North not acted thus . . . there would have been no Irish Republican Army, no I.R.A." But, even to this armed challenge to the then British-certified territorial integrity of Ireland, the Irish were slow to respond. Not until four years later, on Easter day 1916, did the first major proponent of Irish nationalism, Padhraic Pearse, call for the end of English rule. Few Irishmen listened. But the English government was furious: it was one thing for Parliament to *grant* Irish "home rule," but quite another, and a far more dangerous, thing for the Irish to *proclaim* it. In response, the government arrested the tiny group gathered around Pearse. By hanging him, they gave the Irish what the nationalist movement had so far lacked, a martyr with whom they could relate emotionally.

Prime Minister David Lloyd George then compounded this gift to Irish nationalism by announcing in 1918, in the final desperate phase of the First World War when almost a whole generation of Englishmen had died in the trenches, that he intended to impose military conscription on the Irish. His announcement was met with such widespread demonstrations of refusal that he rescinded the order. His backing down convinced many Irishmen that Britain could be successfully deterred, and on this euphoric note many young men who would have gone into the army instead joined what remained of Pearse's movement, the Irish Republican Brotherhood (IRB). Angered by protests mounted by this group, the government again reversed course and arrested two IRB leaders, Éamon de Valera and Arthur Griffith. The third leader, Michael Collins, who paradoxically was the most militant of the three, was left free, and he began to organize the first true resistance movement. Under the title Sinn Féin (Gaelic for "Ourselves Alone") was created a small body of volunteers who rather grandly styled themselves the Irish Republican Army (IRA).

The IRA had an inauspicious beginning and remained small. The fighting men were ill equipped. At first they had only shotguns and such arms as they could steal from the British. But they had enough to engage in several small-scale provocations. These frightened the British, who declared Sinn Féin and its elected assembly, the Dail Eireann, illegal. At that point, January 21, 1919, what became known as the first Anglo-Irish war began.

But what began was not really "war." Because the Irish combatants were few and ill-armed, they could not engage as guerrillas, so they fell back on the weapon of the weak: terrorism. Their use of that tactic convinced Lloyd George that the IRA was not a political movement but "a small body of assassins, a real murder gang . . ." His characterization was typical of government views of insurgent movements: they are merely "bandits."

As yet, the Irish insurgents were amateurs, without a plan and guided only by instinct and opportunity. But as each attack succeeded, the next became more audacious. Soon they were attacking army and police barracks, courthouses and tax collectors' offices. To prevent the spread of information on themselves, they also robbed the mails and infiltrated agents into the police intelligence offices. Even more significant, they attacked their main local enemy, the Royal Irish Constabulary, blowing up barracks and killing troopers. During 1920 they killed 176 policemen and wounded 251. The tactic paid off. Recruitment to the constabulary fell off dramatically, and some 2,000 men resigned. Then, having disrupted the British administration, the Irish, like most other insurgent organizations, began to replace it. Although it was rudimentary, they created their own organizations to carry out public services wherever they gained the upper hand. These daring actions electrified the previously passive general population so that, as British intelligence concluded, they were soon supported by at least one hundred thousand noncombatants.

The British grew increasingly alarmed. Their paramilitary police, the ten-thousand-man-strong Royal Irish Constabulary, had proved ineffectual. So the British took the combat to a new level. They recruited demobilized English soldiers to act as constables. These men became known derisively as the "Black and Tans"—according to some after a pack of County Limerick foxhounds and others after their khaki uniforms and their Sam Browne belts. Humor, always a long suit in Ireland, also became a psychological weapon. But humor was no protection against armed men. Some of these men and others from the Royal Irish Constabulary were formed into "shock companies" whose members were sanctioned to arrest and imprison civilians without charge. Unrestrained and unsupervised, they often engaged in torture and murder. Inevitably they generated hatred and made recruitment to the IRA almost automatic among young Irishmen.

As the number of Irish combatants grew and as incidents spread and became more spectacular, the British responded by increasing their military force. By May 1921 Britain had fifty thousand soldiers in Ireland, and the British cabinet was considering adding another one hundred thousand. What stayed the cabinet ministers' hand in Ireland was what also shaped British policy elsewhere in their colonies—the realization that the English public would not tolerate an expensive new war. So in a replay of history, the English moved to partition Ireland into the "Englysshe Pale" and the "Great Irishry." The modern equivalent of the Englysshe Pale had been moved by King James I from the area around Dublin to the six northern counties known as Ulster; so the logical move was to split it off from southern Ireland's "Great Irishry." The Ulster loyalists were given a separate government under a regional assembly that became known as Stormont; then the ungovernable Irish in the other counties of the island were constituted as the "Irish Free State." The leaders of Sinn Féin accepted, and the deal was signed in December 1921.

To the English, the familiar adage that "half a loaf is better than no bread" seemed sensible, even statesmanlike, and it should have satisfied everyone and brought peace. But it failed to end the struggle as, fifteen years later, in 1936, it would fail in the Palestine Mandate. Understanding why it gave a reprieve but not a solution is crucial to appreciating not only Irish events but insurgency in general: even a partial accommodation to the demands of insurgents tends to split them into mutually hostile factions composed of those willing to settle for half a loaf and those determined to get it all. Or, put another way, the more or less full objectives of some are met while the unnegotiable demand of others is denied. The southern Irish generally saw the deal as meeting most of their demands, while the nationalists in Ulster saw it as a disaster. This division would shape Irish politics down to our times.

Even for southerners, what the deal amounted to was unclear. What the British had intended was that the Irish Free State would become a dominion within the empire. For some, perhaps then even a majority of the southern population, such a status was acceptable, but for their leaders, including the future president Éamon de Valera, it was not. They insisted that at least their piece of Ireland be independent. So the status of the Irish Free State would remain ambiguous. What the British treated as a dominion, the Irish proclaimed to be an independent state. Three years later, rechristened Eire, it was admitted to the League of Nations.

Looking across the border at the southerners who were able to engage in peaceful and constructive political action, the nationalists of the north grew increasingly sullen and bitter. For them, there was no ambiguity: British rule was firm and clear. Their opposite numbers, the Protestant descendants of the Scots and Englishmen who had arrived in Ireland four centuries before, were alarmed and furious. Believing that they had lost a large piece of what they regarded as their patrimony, they were determined not to accept any further loss. They did not believe they had to. They had grown into a significant community, were armed, and they despised the Irish Catholics.

The Protestants had many advantages—a higher standard of living, better education, and an effective political consensus—but they had one serious disadvantage: they were not a "nation." Like their counterparts in Kenya and Rhodesia, they thought of themselves as Englishmen. So, while they made efforts to perpetuate their status, they were always dependent upon political parties and statesmen in England whom they feared would sell them out.

The Catholic community was the mirror image of the Protestants. Weak, poor, and ill-organized, it had no external supporter. Its cause was no longer central to the satisfied people of Eire. If the Northern Catholics were in any doubt, Eire's prime minister, himself a former leader of the independence movement, Éamon de Valera, declared the champions of the movement for Irish unity and independence, the IRA, illegal in 1936. So if the Irish of the North were to strive for their objectives, they had to act for themselves. For them, Sinn Féin was more than a slogan: "themselves alone" was a reality.

Acting for themselves would be a long and painful process in Northern Ireland as it generally was throughout the colonial world. The first step was to realize what being alone meant. It meant essentially that they had to become more than a collection of dissatisfied individuals: before they could struggle for independence, they had to become a coherent group, a nation. Ironically but understandably, this task was easier in Ulster than it had been in what became Eire. Eire was Ireland; Ulster was not. An Irishman in Eire relaxed among his own people; in Ulster he shared with all Gaelic-speaking

Catholics a sense of difference from English-speaking Protestant "foreigners."

In Ulster, the foreigners—longtime inhabitants though they were—were immediately evident even though they were not so obvious as color made the whites in Africa and India. They controlled the economy, the school system, the government, and the police, and they had their own militia. And they asserted their foreignness: they were determined not to have Ulster be confused with neighboring Eire or themselves mingled with Irishmen. In their eyes, Irishmen were dirty, despicable, drunken, and dangerous. Words like "Paddy," "Muck-Savage," or "Bog-Wog" came easily and often to their lips. They could not wholly avoid the Irish, but they created as complete apartheid as could be managed. Belfast was as segregated as the Arab and Jewish areas of Jerusalem. In every daily contact, the Ulster English strove to impress their superiority and alienness upon the Catholic Irish.

Consequently, the Ulster Irish could not have avoided becoming a coherent group even if they tried to assimilate into the English community. For them, apartheid was an inflexible mold. The accounts of the men who became the IRA and successor groups are filled with anger over the slurs that were their daily experience. They accepted the Anglo-Irish assertion that they were English and hated them as foreigners. Even more, they reacted to attacks upon the Catholic community by Protestant terrorists, the Royal Ulster Constabulary, and uniformed British troops. Or, as they believed, and as we now know, by all of them acting together.

Both sides engaged in armed actions. Each regarded the other as illegal and its actions as terrorism. Repression was provoked by violence and violence answered repression. Belfast ceased to be a city and became a collection of mutually hostile neighborhoods, among which movement was dangerous. Thus, in an almost mechanical sense, the militants on each side forced not only their opponents but the general public into coherent groups that shared anger and fear. Each violent act changed the economic and social climate. As movement was restricted, jobs became scarcer; the standard of living fell; people's lives were disrupted; and family, neighbors, and friends got hurt or killed. Many people, probably even a majority, found the price of the insurgency too high to continue to be paid; some emigrated; others, even some Catholics, went further and began, actively, to aid the forces of law and order, even to inform on their supposed champions. In short, many people got tired. But enough people in Ulster either chose to or were forced to support the militants that the insurgency survived.

Meanwhile, in newly independent Eire, from which the militants hoped for aid and comfort, support almost ceased. The Church turned against them; the population ceased to regard their cause as its cause; even the hero of the Sinn Féin movement, Éamon de Valera, tried to suppress the IRA when it tried to operate from Eire. And, anti-British nationalists though de Valera and his associates had always proclaimed themselves to be, they began during the Second World War to intern IRA members to prevent them from continuing their struggle against Britain.

Repression, of course, also affects the militants. Some are killed and others jailed. But even those who remain free often tire under the exhausting strains of clandestine struggle. As they do, younger men and women, drawn into the vortex of revolt by the repressive policies of the dominant power and by the rhetoric and actions of the militants, replace them and return to the radical policies of the leaders' youth. In Ulster, in the years that followed the end of the Second World War, split after split followed: the IRA were the first radicals; then, in 1969, their organization sundered. The old guard, who became known as the Official IRA (OIRA) began to think that what Ulster needed was not warfare but socialism; that it should be possible to "uplift" the Irish people within the context set by the British. The more militant insurgents disagreed so strongly that they created a new organization known as the Provisional IRA (the PIRA, or "Provos"). The Provos saw their central task to be fighting the Protestant paramilitary gangs that were harassing northern Catholics. As one of their leaders put it, "We receive our support from the Nationalist people and it is our job to defend them." That task was something the OIRA not only ceased to want to do but also ceased to be able to do, since on the eve of the split in 1969, their "army" was believed to have had only ten guns. In recent times, a new group, called the Real IRA, has arisen as the Provos themselves have moved toward less violent means of action. In the view of the Real IRA, anything less than a fully independent and unified Ireland is unacceptable, and those who accepted less are traitors to the cause. And even the Real IRA did not satisfy the desire of some militants for action so a group split off from it, calling themselves Continuity IRA.

The young men and women who became Provos in the 1970s—and the more recent Real IRA—thus reverted to the program of their IRA fathers or grandfathers in the years after the First World War: it was to treat Ireland as the single unit proclaimed in Dublin in 1916 rather than the dual territory established by the Anglo-Irish treaty of 1921. In their eyes, the Stormont government the British had created in the Government of Ireland Act thirty years before for the six northern countries was merely a subterfuge; as long as it existed, the British could pretend that "Ulster" had already achieved quasi-independence. To convince the Irish that this was not so and that their fight was "with the English for our God-given right to nationhood," the Provos believed they had to make the Stormont government fail and thus force the British to "come out into the open" as the real—"colonialist"—rulers of the North.

To achieve this objective, the Provos began a ruthless and widespread campaign of terror. Their new weapon was the car bomb, which would become a standard weapon of militant movements elsewhere. But they also used simple bombs, mortars, machine guns, and even pistols—any weapons they could acquire—wherever they found targets. They found many. During the years immediately after their formation, the Provos carried out an average of seventeen shooting incidents and nearly four bombings a day. Among their victims were young men, mainly Protestant, who joined or were likely to join the Royal Ulster Constabulary or the Ulster Defence Regiment, of whom they wounded more than a thousand and killed fifty-eight.

Their attacks were intended to be and were terrifying. Even

though they actually caused fewer deaths than accidents on Ulster's highways, they were far more dramatic. In turn, the British began a process of interning men and women suspected of involvement. Ultimately, there would be nearly four thousand in prisons and makeshift camps. Just as the internment campaign began, the Irish population of Derry held a rally on January 30, 1971, to protest this new round in the troubles; as they gathered, British troops fired into the crowd, killing fourteen people. This event was widely condemned after an investigation by the "Insight Team" of reporters from the English newspaper, the *Sunday Times*, and was enshrined in the Irish mythology as "Bloody Sunday."

Terrorism thus provoked actions that drew supporters to the nationalists, but also, because it disrupted lives and killed bystanders, it lost the Provos the support or acquiescence of increasing numbers of the Northern Irish population. Recognizing that their tactics were eroding their support and would be fatal to their movement, the Provos decided to offer a plan to end the conflict. The Eire Nua (New Ireland) for which they were struggling, they proclaimed, would have a regional parliament, the Dail Uladh, in which the resident Protestants would have a significant representation. To get to that objective, they demanded five concessions from the British government, which amounted to recognition of Ulster as a quasi-autonomous part of Eire. To give the British a chance to acquiesce, they declared a truce but coupled it with the implicit threat to intensify the war if the British did not agree. When nothing happened, they modified their demands, and on March 10, 1972, declared another truce. The British suspended the northern assembly, Stormont, but refused the other demands. Unsatisfied, the Provos renewed their campaign of terror. To this, the British government responded by arranging a meeting with the leaders of the Provos in London.

The meeting, predictably, was a failure: neither side thought there was anything on which to negotiate. Was this a reasonable assessment? It is a question worth asking not only for Ireland but more generally, as it arises in nearly every insurgency. Theoretically at least, there were possibilities in Ireland such as Britain implemented in other colonies, including withdrawal as in Rhodesia and Kenya. Another was a further division of Ulster along ethnic-religious lines. That division already existed de facto and *could* have been formalized. It could have been implemented politically by allowing predominantly Irish districts to become independent or join Eire and physically by some sort of barrier such as Israel created to fence off the Palestinians. Where ethnic-religious groups were mixed, there probably would have had to be an exchange of populations like the one Greece and Turkey implemented at the end of the First World War. Either of these "solutions" certainly would have been painful and costly and might not have been effective. There *could* have been agreement on a transition period to independence or amalgamation with Eire, during which period British troops *could* have been replaced by a United Nations peacekeeping force as had already become the device of choice in a number of former colonial areas. And there were probably other options, but all would have carried high political prices. Both the Provos and the British thought they could win, and neither thought the other a suitable counterpart for negotiation. Consequently, the fight continued and intensified.

The failure of negotiations and the massive increase of insecurity convinced the British that they had to intervene directly. The Provos regarded this decision as a strategic victory because they wanted to draw the British deeply into the conflict in order to keep the war painful for the English public and so bring about a complete British withdrawal. They also hoped, like the Algerians, South Africans, Palestinians, and others, that world opinion would become mobilized against their opponent. But the Provos realized that militarily they, and particularly their civilian supporters, would pay a heavy price. They did.

With a military force of only about a thousand men and women, the Provos could not hope to defeat the British expeditionary force, which, with its Royal Ulster Constabulary and Ulster Defence Regiment auxiliaries, numbered nearly thirty thousand. Almost any concentration of Provos would draw down on them overwhelming numbers and the much feared combination of helicopters and armored cars; so in general, the Provos were unable to engage in guerrilla warfare but were restricted to acts of terrorism carried out by individuals or small groups. One Provo picked up the familiar metaphor, describing the campaign as "the war of the flea," harrying the elephantine British army in a war of attrition, harming British financial interests, making Northern Ireland ungovernable while gaining support from the world community.

Against the flea, the use of massive armed force appears attractive to the occupying power. Just as the British believed in the American insurgency, the French in the Spanish resistance, and the Germans in Yugoslavia, numbers and equipment appeared to guarantee success. The "elephant" seemed certain to win. But the very military power of the dominant government proved in Ireland, as it often did elsewhere, self-defeating. It actually enhanced the strength of the militants because it involved violent methods that were repugnant to the general public and seemed to prove the militants' propaganda.

Put another way, the dominant power and the insurgents were fighting overlapping but different wars. The dominant power aimed to destroy the insurgent movement while the insurgents aimed to dishearten the occupying power and convince it that the struggle was too expensive to maintain. The dominant power relied on force and avoided political action, while militants, having limited force, sought to make the struggle political but also violent. Each side pushed its own agenda. The British augmented their armed forces while the Provos accentuated, as they spelled out in their guide to their overall strategy, their "Green Book," the use of their only effective weapon, terrorism.

In 1974 the IRA began a prolonged and spectacular bombing campaign in England. Included were a variety of targets from pubs to Parliament, from the huge Canary Wharf complex in the center of London to Harrod's vast department store, from the Brighton hotel in which Prime Minister Margaret Thatcher was meeting the leaders of the Conservative Party to the boat in which the Earl of Mountbatten was riding toward his Irish estate, and from the cabinet at Downing Street to Heathrow airport. In riposte, the British authorities determined to emphasize the IRA/Provos' criminality by removing captured fighters from political prisoner status. Since this definition of their role was the very essence of their concept of their struggle, the prisoners resisted. They refused to wear prison uniforms or to abide by prison regulations. Finally, in October 1980, a number of the prisoners, insisting that they were prisoners of war, began a hunger strike. The most famous of these prisoners was Bobby Sands who, in the midst of this strike, was put up for a seat in Parliament. He was elected but died during his hunger strike; when that happened, one hundred thousand people attended his funeral. The Ulster Irish found it to be one of the most moving episodes in their history. Whatever else it did, it fixed the image of the IRA as the national liberation movement.

But the war continued. By August 1989, British troops had been engaged in the Irish war for twenty years. Periods of intense violence were interspersed with periods of relative calm. Then an attempt was made to return to the negotiating table. Since the two main sides had not been able even to agree on what negotiation meant, an effort was made to use an external arbitrator. The skillful work of U.S. senator George Mitchell resulted in the Good Friday Agreement of April 10, 1998. The agreement was opposed by the Protestants as giving too much to the Catholics, and it hung in abeyance as the Protestants began a campaign rather like the Colon (settler) Secret Army Organization in Algeria, using terror to try to prevent the home government from compromising with the insurgents. So the Protestants and Catholics switched sides at least in their tactics: as the Protestants went underground and began a campaign of terror, the IRA came into the open. In this movement, a former active militant of the Provos, Gerry Adams, came out of the shadows to play a new role as the political leader of the Provisional IRA.

After years of misunderstandings, fears of treachery, and lapses into violence, on July 28, 2005, the leadership of "Óglaigh n hÉireann" announced that it had "formally ordered an end to the armed campaign." It proclaimed that "all IRA units have been ordered to dump arms" and that "all volunteers have been instructed to assist the development of purely political and democratic programmes through exclusively peaceful means . . . We believe there is now an alternative way to achieve this [implementation of the Good Friday Agreement] . . . including our goal of a united Ireland and to end British rule in our country."

Some saw this as the end; others refused to heed the order. So fighting within each community, between communities, and between both and Great Britain continued. As the British press reported on February 2, 2006, despite its announcement that it had decommissioned its weapons and agreed to disband, the "Independent Monitoring Commission" reported that "the IRA had not disbanded" and continued to prepare to launch a new round of insurgency. During 2006, a number of breakaway members of the Provisional IRA, itself a dissident group of the IRA, calling themselves the Real IRA, were arrested by police in actions in Spain, France, and Ireland.

In January 2007, the leader of Sinn Féin, Gerry Adams, asked his followers to begin to cooperate with the newly formed Northern Ireland police force that replaced the Royal Ulster Constabulary and in March reached a deal—proclaimed in his first-ever meeting with the leader of the Protestant Unionists, Ian Paisley—to form a government of Ulster. The Unionist–Sinn Féin deal was signed on May 8, 2007. Both sides were shocked by the realization that their conflict had cost some 3,700 deaths and untold misery for their followers.

Will it be, can it be, the end of the insurgency before Ireland achieves a basic unity? Arguably, the Irish insurgents have a choice. Some insurgents do not. For one of the most spectacular of those without a choice, I turn next to the resistance mounted by Tito's "Partisans" in Yugoslavia against the Germans during the Second World War.

CHAPTER 5

TITO AND THE YUGOSLAV PARTISANS

ON THE DISPUTED BALKAN FRONTIER OF THE GREAT EMPIRES OF THE Austrians, the Russians, and the Ottoman Turks, the "Land of the South Slavs"—Yugoslavia—has been invaded and settled, ravaged and rebuilt, raped and converted time after doleful time for centuries. Each of the great powers has left its impress so that the sum of its history is that it did not become a nation but a collection of nations. Religiously, its inhabitants were divided among Orthodox and Catholic Christians, Sunni and Shia Muslims, Jews and Gypsies. Ethnically, most of the population was made up of Slavs, but they shattered along the fault lines of language, geography, and custom into Western-influenced Catholic Croats in the north who read in the Latin script and Russian-influenced Orthodox Serbs in the center who read in the Cyrillic script. In addition, clusters of Hungarians (or Magyars), Italians, Albanians, Austrians, Bulgarians, Czechs, Germans, and Turks lived in little congregations that dotted Yugoslavia's wide plains and clung to its lush Adriatic coast. In the frequent times of danger, its rugged mountains and high valleys have provided refuge for heresies of each of the religions and for dissidents from the ruling establishments. In the tragic events of the 1990s, we have seen how deep and how virulent are the suspicions and hatreds that have long divided these close neighbors.

Influenced by the tide of nationalism that in the nineteenth century was sweeping through the Ottoman and Austro-Hungarian empires, Serbia, Bulgaria, Montenegro, and Greece formed themselves into states from the wrecks of these empires. They drove the Ottoman Turks out of the Balkans in 1912 and then fell upon one another over the division of the newly liberated territories. In its grand but futile effort toward a more peaceful world at the end of the First World War, the Paris peace conference allotted most of what later became Yugoslavia to the "United Kingdom of Serbs, Croats and Slovenes."

The very name of the new kingdom was a sign of its fundamental problem: it wasn't united. The first appointed king, a Serb, actually opposed the creation of the new state, so he was deposed. His successor, King Alexander of Serbia, struggled to transform the disparate and mutually hostile communities under his rule by imposing upon them a highly centralized state that, in 1929, he renamed Kraljevina Jugoslavija (the Kingdom of Yugoslavia). In his attempt to break down the barriers dividing the inhabitants, he created a state as totalitarian as Mussolini's Fascist Italy, Hitler's Nazi Germany, or Stalin's Communist Soviet Union. As one historian has written, "He made the laws, commanded the army, nominated all officials, declared war, and made peace. He was irremovable. Simultaneously, he declared the end of freedom of person, of press, of speech, of assembly. A death sentence was prescribed for any attempt to overthrow the government, for spreading propaganda aimed at the existing social order, and for relations with any revolutionary organizations abroad . . . Police brutality and the arrest of political opponents were standard features of the royal dictatorship."

Yet, all in vain. The component parts of Alexander's kingdom sullenly guarded their separate identities and actively nurtured their mutual hatreds; most remained poor, illiterate, bellicose, and rebellious. One of the strongest, most active, and most violent groups, known as the Ustase (the "rebels"), assassinated the king in 1934. They had not aimed at a more open or democratic government, and Yugoslavia did not get one: Alexander's successors continued along the authoritarian lines he had laid out. Under them, indeed, the police state became even harsher as it forged closer economic and political ties with Fascist Italy and Nazi Germany. The logical outcome was reached on March 25, 1941, when the Yugoslav government signed the Axis pact. That was too much even for the ultra-conservative but strongly nationalistic Yugoslav officer corps: it overthrew the government in a coup d'état, deposed the regent who had taken over from Alexander, and installed Alexander's son Peter as king.

Yugoslavs of all persuasions were delirious with joy, but their joy was occasioned by conflicting expectations. Each was to be disappointed. The only action on which they agreed was attacking the resident Germans and looting their shops. The reaction abroad was immediate, understandable, and divided. Britain, the United States, and the Soviet Union were delighted, while the Germans were furious. Hitler regarded the coup as a personal insult. He was particularly annoyed because it came just as he was showing the foreign minister of his new Japanese ally how powerful his Germany had become. But Hitler realized that he could not leave the vengeance he determined to take to his fumbling ally Mussolini, who had recently received a bloody nose in his attempt to invade Greece, so he ordered the German army to punish the Yugoslavs. By the middle of April, with shock and awe, a Wehrmacht blitzkrieg had sliced through the Yugoslav army and was on the way toward Greece.

In twelve days the Yugoslav war was over. Hitler thought his mission was accomplished. Yugoslavia lay prostrate: its army melted away, and the government fled. Whatever remained was his. So he immediately began to dismember the corpse of Kraljevina Jugoslavija. To Germany he gave most of the Slovene area adjacent to Austria. For Germany the land was desirable but not the people. So German troops and the Gestapo immediately began wiping off all signs of the people. They burned books and removed inscriptions even from graves that were in Slovene language. Next they packed some sixty thousand Slovene villagers into freight cars to be "dumped" in the occupied Serb areas; and, finally, they turned over the then vacant houses and farms to German settlers. Where Germans were already settled, in a small enclave on the Rumanian frontier, Hitler created what was virtually an extension of the German state, whose inhabitants showed their gratitude by providing the Wehrmacht with an SS division. Imperiously, Hitler allotted the Adriatic coastal area and most of the Adriatic islands south from Trieste to the Albanian frontier to Italy and Italian-occupied Albania, while dividing Macedonia between Albania and Bulgaria. Hungary got what was left.

Having split them apart politically, the Germans moved to accentuate the Yugoslavs' traditional animosities. Croatia received relatively lenient treatment, which made the others hate it. It was allowed a puppet government under the Ustase, which was led by a Croat who modeled himself on the German *fuehrer* and the Italian *duce*. Both gave him every encouragement; even the pope received him, and the resident Catholic hierarchy lent him both support and blessing. For the Germans, he was an ideal surrogate. So they spent considerable effort to win over the Croats; in large part, they succeeded. The puppet government declared war on the Allies and deflected popular hostility away from the Germans toward local scapegoats; it hurled the troops and paramilitary commandos the Germans had allowed it to keep into ethnic cleansing. Their targets were the relatively defenseless Serbs, Jews, Muslims, Gypsies, and other minorities in the Croatian zone. Tens of thousands were massacred.

Meanwhile, the Germans treated the Serbs in "Serbia" as defeated enemies. The contrast to their relative favoritism of the Croats and their barbarity to the Serbs had the desired effect of promoting ethnic hatreds. At every opportunity, the Germans magnified this rift. They hardly needed to do so: Croat Ustase gangs were more than willing to help. But in the interest of economy, since the regular German forces were needed to fight the British in Greece and to attack the Russians, the Germans put what they now defined as Serbia under a puppet government led by a former minister in the royalist regime. This government was assigned the task of rounding up Serb men and women to be shipped to Germany to perform slave labor.

Terrified of both the Croat gangs and the Serb headhunters, small groups of Serbs fled to the mountains or forests. There, at first, they were just hunted refugees, leaderless and out of touch with any organization. The resistance did not yet exist, nor, even if it had, could its members have maintained contact, which probably none would have wished to do, with the royal government-in-exile in London. They were essentially on their own.

How these various German, Italian, and Croat moves created a climate for insurgency is obvious: Yugoslavs of the various communities and classes had felt exploited and deprived before the arrival of the Germans and Italians; they had little love for the then exiled royal government; but they quickly came to hate the foreigners and their puppets who now ruled them. Their anger was accentuated by the further losses that the invasion inevitably entailed. Not only were lives disrupted, property damaged or destroyed, and people wounded or killed, but the incoming foreigners ran roughshod over the arrangements and traditions to which each community was accustomed. The German occupation policy, echoed in part by the Italians, virtually forced the more deprived and more frightened parts of the population to resist.

What initially set the Serbs apart from the Croats was that, in their misery and fear, isolated individuals and small groups began to steal from German supply dumps and kill German soldiers. Probably their first violent acts arose from their desperate search for food. To the Germans, however, they were acts of sedition. In retaliation, the Germans executed hundreds of hostages. Each retaliation fueled further hatred. Impotence turned to rage. And rage provoked what the Germans saw as tactically infuriating but, strategically, essentially meaningless "bandit" raids. Although the population was willing to fight and was partially armed (since at least some soldiers had kept their arms when they fled the German blitzkrieg), they had lost all military organization and were leaderless. The Germans thought they could track them down and kill them. Instead, the tempo of attacks increased. Into this cauldron came two opposing groups.

The Cetniks (the word comes from the Serbian *ceta*, which has almost exactly the same sense as the original Spanish meaning of *guerrilla*, "little war"; the C is pronounced as "ch") were mobilized and led by Drazha Mihailovic, a Serb who had been a colonel in the defeated royalist army. He did not have much to work with. He began the Cetniks with only twenty-six like-minded former officers and soldiers. His adherents multiplied as a result of German and Croat tyranny, but his force was always small and sometimes virtually disappeared in the holocaust unleashed upon them by the Germans and Italians.

A man of the extreme right, Mihailovic had no interest in or sympathy for the social or economic grievances of the Serbs. Fervently religious and strongly anti-Communist, he was less disturbed by the presence of the Germans than by the potential threat of the Russians. And, as a cautious staff officer, he saw his role as preserving his limited forces for the day when the British would drive out the Germans.

Mihailovic had reason to be cautious: having learned that in October 1941 the Germans had massacred about seven thousand Serbs, he feared that, if provoked, they would carry out reprisals on a scale—they had publicized that they would kill three hundred Serbs for every German soldier lost—that would virtually wipe out the Serbian people. But his military caution and his social conservatism were not winning formulas for wartime Yugoslavia. So, although he was first in the field and had the advantage of being recognized by both the British and the Soviet governments as *the* leader of the insurgency, he quickly lost the initiative. Then, pressed on the one side by the Germans and on the other by the newly organizing Partisans, and without effective support from either the royal government-in-exile or the Allies, he lapsed first into passivity and then into collaboration with the Germans and Italians. His decline and fall as an insurgent began on November 10, 1941, when he met with a German intelligence officer to discuss possible collaboration. Nothing came of that meeting, but a few months later, in March 1942, one of his officers worked out what was virtually a mutual assistance treaty with the Italians to attack other insurgents in return for supplies. Probably as a result of these moves, many of his followers deserted him to join the Partisans.

The Partisans were the creation of one of the most remarkable men of the twentieth century, Josip Broz, whom the whole world knows as Tito. Tito had the raw materials from which to manufacture insurgency at hand: all over Yugoslavia small groups of frightened, angry, and armed men and women were hiding out in the forests and remote valleys. They thought they faced almost certain death if they could not find ways to defend themselves. Both Mihailovic and Tito offered them ways to fight the Germans; militarily, at least in concept, there was little to distinguish their organizations. But Tito also captured the *political* element of insurgency in a way that Mihailovic did not even comprehend. While Mihailovic was a man of the old regime, Tito stood for a program of social revolution. As a result his armed "nation-in-motion" grew to perhaps 150,000 by the fall of 1942; of these, perhaps 20,000 were actual fighting troops.

The beginnings of the Partisans were almost as meager as those of the Cetniks. But the growth of the Partisans was faster and more enduring. As described by Churchill's personal representative to the Partisans, Brigadier Sir Fitzroy MacLean, who had unparalleled access to Tito, the Partisan leadership, and British intelligence officers, later wrote, "within two or three weeks of Hitler's attack on the Soviet Union and the party's call to arms, a number of guerrilla bands were operating in Serbia and elsewhere under Communist command. They consisted of small groups of determined men and women, mostly party members, who, at a word from party headquarters, had taken to the woods and forests, equipped with such arms as they had been able to collect, with cudgels and axes, with old sporting guns and with anything else they could lay hands on. For further supplies they depended on what they could capture from the enemy and on what the country people would give them."

Momentum quickly built up. MacLean again: "Every day petrol and ammunition dumps were blown up, convoys ambushed, trains derailed, enemy outposts raided and more and more arms and ammunition captured. Soon large areas of the country were in the hands of the insurgents, while in the towns the German garrisons lived in constant fear of attack."

The German Staatsrat (or privy counselor, who served as a sort of inspector general in Yugoslavia) painted a gloomy picture of the occupation for his bosses in Berlin. It has not proven possible, he wrote, to create a "minority question . . . among the Serbs, as it was with such success among the Croats . . . they follow the Communist bandits blindly. With their slogans the Communists have succeeded in rallying round them elements who in the past would never have dreamt of co-operating with them. Some go so far as to prefer Bolshevism to occupation by our troops—and these are people on whose co-operation we were counting. Only one means is left: armed force . . . My impression is that even the news of the capitulation of the Soviet Union would not cause these bandits to capitulate. They are tougher than anything you can imagine. What is more, their organization is excellent. It might serve as the classical example of a perfect secret organization."

What Tito had done was to forge from disparate and mutually hostile elements of the Yugoslav population a coherent movement. As we have seen in previous insurgencies, including the American War of Independence, such overcoming of internal disagreement, often amounting to civil war, is often brought about by terrorism: the group that ultimately wins does so because it immobilizes, drives away, or kills its opponents and galvanizes or frightens the uncommitted portion of the population into support or at least acquiescence. But Tito did not need to conduct a terrorist operation to impress upon his followers a sense of solidarity with one another and opposition to others: the Germans, Italians, Bulgarians, and the Croatian Ustase did that for him. As MacLean wrote, "The whole of Bosnia ran with blood. Bands of Ustase roamed the countryside with knives, bludgeons and machine guns, slaughtering Serbian men, women and little children, desecrating Serbian churches, murdering Serbian priests, laying waste Serbian villages, torturing, raping, burning, drowning. Killing became a cult, an obsession . . . Some Ustase collected the eyes of the Serbs they had killed . . . proudly displaying them and other human organs in the cafés of Zagreb. Even their German and Italian allies were dismayed by their excesses"—as well they should have been, because, terrified and infuriated, those who survived became recruits for the insurgency.

The raw materials of insurgency were thus provided for Tito, but it was Tito who hammered them into an organization that could survive the onslaught of eight German divisions armed with artillery, tanks, and aircraft, outnumbering his forces at least five to one, and win. So we must ask how he did it. I begin with a summary of what is known to have shaped the man who appears on the world stage only when already fully formed.

Born Josip Broz in the Croatian village of Kumrovec, he was the seventh of fifteen children. His father owned a small farm and supplemented his meager living by carting supplies for his neighbors. Josip left school after only four years when he was twelve. He got his first job at fifteen and then became an apprentice to an ironmonger. Moving on, he worked in Zagreb as a mechanic, a field that continued to fascinate him throughout his life, and there had his first taste of politics organizing for the Metalworkers' Union and participating in activities of the local Social Democratic Party. When the First World War broke out, he was called up and fought on the Carpathian front against the Russian army. There he learned the rudiments of warfare and became a platoon commander; more interestingly and probably more to the point given his later career, he seems to have distinguished himself in the sort of bold, behind-the-lines forays that the Partisans would later employ. Another experience may also have been seminal: getting pinned down by a Russian attack, he was severely wounded and captured. Later, perhaps remembering this painful event, he would do all he could to prevent his Partisans from ever getting caught in static combat. Carried off to Russia as a wounded captive, he spent a year in a hospital—nearly dying of a combination of his wound, pneumonia, and typhus—before being put to work helping to build the Trans-Siberian railroad in the Urals. That was where he learned of the first (February 1917) Russian Revolution.

The Revolution fired his imagination, and he quickly identified with the most radical faction, the Bolsheviks. After a series of narrow escapes from the White armies, the political police, and roving bands of outlaws, always on the run, he was caught trying to leave Russia and was put into prison in St. Petersburg. It was in St. Petersburg that he witnessed the October Revolution that brought the Bolsheviks to power. As the first British officer who parachuted into Tito's band, F. W. D. Deakin, who shared his battles and retreats and got to know him well, described the effect of those events on Tito, "He had with his own eyes witnessed the triumph of the Bolshevik Revolution and the establishment of the Soviet State. He had seen the working class, to which he himself belonged, rise and seize power. He had seen, or thought he had seen, the Future—the Future for Russia and for the whole world. He had found a cause by comparison with which family, religion, fatherland counted as nothing, a cause which demanded complete and absolute devotion. He had become a Communist. Henceforward, he had but one aim: to spread the gospel of Communism, to bring about as soon as possible a Communist revolution in his own country."

Returning to Yugoslavia, Tito/Broz threw himself into Communist Party activities while working in a shipyard, and then in a factory. Often arrested, he used each stay in prison to make up for his lack of formal education by reading everything he could find in Croatian, Serbian, Russian, German, and French and by participating in seminars with fellow prisoners on Marxism. Those times, he later told Deakin, were "just like being at a university." In the intervals between imprisonments, he gradually worked his way up to a responsible position in the small, embattled, and then waning Communist Party.

By 1935 the Yugoslav political police had virtually destroyed the Communist Party, imprisoning, exiling, or killing most of the senior members. The police were closing in on him as well, so his colleagues decided Tito/Broz must get out of Yugoslavia. He returned to Russia to join the staff of the Comintern in Moscow. He later admitted that what he had seen of Communism there disappointed him, but he did not allow that impression to affect his loyalty. His loyalty must have been evident to his handlers because, after a year, he was sent back to Yugoslavia, in disguise, and assigned the task of recruiting volunteers for the Spanish Republican Army. Very successful at this, he enrolled about fifteen hundred young men, some of whom would later serve with the Partisans.

At the height of Stalin's purges in 1937, he was called back to Moscow—a dreaded trip that was then often just one-way for foreign Communist leaders. It was not for him. He was actually promoted and put in charge of the Yugoslav Party. Returning to Yugoslavia, Comrade Walter, as Tito/Broz then became known, threw himself into the task of rebuilding the Yugoslav branch of the party and making it useful to Russia. His task, as he later said, was to make the Russians popular in Yugoslavia: "It was our principal activity; it commanded the bulk of our funds."

He carefully followed the party line. When Stalin tried to make a deal with Hitler and even when Germany invaded Yugoslavia on April 6, 1941, he did not react. During that period, in a move that must have disturbed even a completely dedicated Yugoslav Communist like Tito, Stalin withdrew recognition of the Yugoslav Party in his attempt to woo the Germans. Comrade Walter kept mum. It was not until the Germans attacked Russia on June 22 that he led the Yugoslav Communists against them. Then he reacted with what must have been pent-up energy, frustration, and anger arising from that period of Russian opportunism.

Himself a Croat, Tito particularly courted the Serbs, but in general he sought to rise above the religious, ethnic, and cultural divisions of the Yugoslavs. Astonishingly, he began not only by organizing his own forces but by trying to win over the royalist Cetniks under Mihailovic. Already in the middle of September 1941, he met with Mihailovic and proposed joint operations. Mihailovic refused. So Tito tried again a month later, this time sweetening the deal by giving Mihailovic four hundred rifles. This gift, turned out by a factory the Partisans had seized, was also meant, of course, to demonstrate Partisan power. If this gesture was not enough, Tito doubled it by showing how much territory his men controlled. He did this by escorting to Mihailovic's headquarters a British officer and two royal Yugoslav officers, who had been landed from a British submarine. If Mihailovic was impressed, he did not show it in the one issue that mattered to Tito: he again turned down Tito's proposals for joint action. Then, not realizing the shift in their relative power that Tito had tried to demonstrate, Mihailovic followed up the meeting by an attack on Partisan forces. In this attack, the Cetniks were decisively defeated.

Meanwhile, the finally aroused Germans began the first of what would be seven major offensives aimed at the Partisans. Annihilation, not defeat, was their aim; the Germans took no prisoners. In this first campaign, Tito and his small staff were nearly caught. Those who *were* caught, including the wounded, were shot. The Germans meant to terrify the Partisans and their supporters, but the results were almost exactly the opposite. On December 1, 1941, they tracked down sixty wounded men and women who had crawled out into the fields surrounding the town they had just captured; most had lost legs or arms and had to be abandoned by the fleeing Partisans. The Germans drove their tanks backward and forward over them until they were all dead. Those Partisans who had escaped the German encirclement because they were able to move rapidly realized that they had to win each battle or suffer a similar fate. There could be no surrender.

What was this army? Was it even an army? Or were the Germans right that it was just a ragtag group of peasant bandits? We can form a remarkable picture of it because the British liaison officer, Captain (later Colonel) William Deakin, was parachuted into Tito's temporary headquarters in the wild mountain country of central Yugoslavia, just as the German army was encircling them in what they called the fifth German offensive. He was to spend months with them, sharing their battles and their bivouacs, their miseries and their meals, watching and listening. This is what he wrote: "In physical appearance it was a motley army, the troops dressed in captured Italian or German uniforms, disconcerting at times in abrupt encounters at close quarters in the woods, in the grey service dress of the former Royal Yugoslav army, or in peasant clothes. They were lightly armed with rifles or light submachine-guns with bandoliers of ammunition and some of them with Italian hand grenades attached to their belts."

Since it is so rare that we get the report of an outsider who was also a professionally trained eyewitness to guerrilla combat, it is worth listening carefully to Deakin. In the first campaign where he was often right beside Tito, he more participated than witnessed. He was wounded and his colleague, Captain William Stuart of British Military Intelligence, was killed. In that action, which was typical of many,

a considerable force of lightly armed and hardened fighters of some twenty thousand men was engaged in a fanatic search to outmaneouvre a highly professional military operation of encirclement, conceived by the German command and conducted in the main by elite S.S. and Alpine troops as part of a force more than five times the number of their opponents, equipped with heavy weapons and mountain artillery, supported by fighter and light bomber aircraft . . .

Partisan tactics were conceived in the counter image of the enemy, and the special style of their actions engendered the miracle of their survival . . . The Partisans were moving on inner lines within a tightening ring. Lightly armed and familiar with the terrain, trained to operate instinctively in small and isolated parties, their units could evade encircling thrusts of the enemy. Skilled in ambush and experts in night fighting at close quarters—for which the Germans showed peculiar reluctance—the Yugoslav troops could often gain brief but vital local superiority. The protection of the general movements of their columns was decisive to the outcome of the battle. This narrowed to a race for the mountain crests, which dominated in scattered dots the river crossings and the upland tracks. Each height was the scene of hand-to-hand clashes without quarter, to be held at all costs in unison with the moving columns of the main group of Tito's forces with their sick and wounded.

The probing of the enemy circle was the supreme test of Partisan tactics. The decisions facing Tito and his Staff were brutal in their simplicity: to mislead the enemy as to the main direction of his forces seeking to break out of the ring, to preserve a striking force with the purpose of completing such an action, and at the same time to protect the columns of the hospitals.

Deakin and other British officers were struck by the flexibility and resilience of Tito's force: when the Germans crashed into an area, they laid low or disappeared; then when the Germans moved on, the guerrillas flowed back into the area "whose link with the Partisan forces had not been broken, and [resumed their control] by an underground structure of local administration which had survived the passing storm." As Tito later explained to Brigadier MacLean, "while hitting the enemy as hard as possible, they must at all costs deny him a target at which to strike back. Manpower was precious. Pitched battles against a stronger and better-equipped enemy must be avoided. Use must be made of a large number of small, highly mobile detachments. These would be closely linked and so be capable of combining to form larger units of battalion strength which could be used as shock troops in an emergency and then split up again into their component parts . . ."

Tito freely borrowed from the Russians, who during World War II also employed partisans against the Germans; however, his military structure was notably different. Whereas the Russians used irregular troops in the classical eighteenth-century mode of *la petite guerre*—as auxiliaries to their main battle formation, sending them behind German lines to gather information, pick off isolated groups, and sabotage facilities—Tito's Yugoslavs were the only army. Correctly put, they were not partisans. They were guerrillas.

Tito's campaign also illustrated the importance of terrain. His guerrilla forces were able to retreat into areas that were difficult for the more heavily armed German and Italian troops. So also had the Spanish guerrillas used their mountains and valleys in their war against Napoleon. And, similarly, the Arab guerrilla force that fought against Ottoman armies in the First World War used the desert into which regular Turkish troops could not follow. However, with the growth of military technology, terrain proved not so important an advantage as has often been thought. During the Second World War, the Germans used mountain troops, even ski troops, and aircraft to penetrate what in previous combat were safe havens. As we shall see, the helicopter gave the Americans in Vietnam and the Russians in Afghanistan even more powerful means of action and mobility. So, while it is important, terrain is not decisive.

Also in looking at guerrilla warfare, observers often exaggerate the importance of foreign aid. As one of Tito's close collaborators, the Sephardic Jewish officer Mosa Pijade, wrote, it was a "fable." A British liaison officer with Tito's forces, Basil Davidson, summed up the aid question succinctly, their "only source of supply was the enemy; for it was not until 1944 that British supplies began to arrive in appreciable quantities." The Russians were no help either. Despite Tito's frequent requests for arms and other supplies, the Russians made no move to supply them; they warned Tito that "technical difficulties are enormous. You should not count on our overcoming them in the near future. Please bear that in mind. Do all you can to try to get arms from the enemy and to make the most economic use of what armament you have." One after another guerrilla forces around the world found that they were virtually alone: Mao's guerrillas in China, the "internal" resistance forces in the Algerian war against France, the Filipinos fighting against the Americans, and the Viet Minh in the early years of their war against France all had to get most of their weapons and ammunition from their enemies.

More important than terrain and outside supply were two other aspects of the Yugoslav resistance. The first of these was administration. As Brigadier MacLean observed, the "nation-in-motion" seized every opportunity to establish "liberated areas," in which the former administration was destroyed. In its place, Tito created *odbors* ("people's councils")—possibly inspired by what he had seen or read about during his visits to Russia, the committees the Russians called soviets—to manage local affairs; the committees opened schools, published a newspaper, and even arranged sports events and entertainment. Although the guerrilla "state" was forced into a nomadic existence by Wehrmacht onslaughts, it stopped long enough to engage in a range of activities. Already in September 1941, it established itself at the town of Uzice, which had a population of about fourteen thousand. There Tito's men found a branch of the national bank whose vault yielded a small fortune in Yugoslav currency; this enabled them to buy rather than simply seizing food and other supplies. This policy endeared them to the peasants, who had nearly always been looted by passing armies. Almost better, the guerrillas found that the local cigarette factory had retained nearly 350 tons of tobacco and twenty-three truckloads of cigarette paper. Since the Yugoslavs were heavy smokers, Tito's men turned these raw materials into "Red Star" cigarettes. Excellent propaganda weapons they were: every time a person smoked, he implicitly recognized the guerrilla regime. Even better were real weapons. The town boasted a small factory that was capable of producing about four hundred rifles a day. (It was one day's production that Tito had grandly given to Mihailovic, and, since the factory stamped each with a red star, it must have been a gift hard for Mihailovic to accept.) Two years later, as Deakin found in a different area, the guerrilla "state" was even running a small railroad and had its own postal service with its red star prominently displayed. As with the lighting of a cigarette, each lick of a stamp affirmed one's relationship to the cause.

Administration and production were crucial to the guerrillas' success, but as MacLean observed:

> [P]erhaps their greatest strength was the idea which inspired them. In guerrilla war ideas matter more than material resources. Few ideas equal revolutionary Communism in its strength, its persistence and its power over the individual. Communism gave the Partisans a singleness of purpose, a ruthless determination, a merciless discipline, without which they could not have survived, let alone succeeded in their object. It helped them to overcome their old national feuds and divergencies. It inspired

in them an absolute devotion to their cause which led them to count as nothing their own lives or the lives of others. It brought them a ready-made intelligence system, a well-tried, widespread underground network. It endowed them with an oracle: the party line. To what had begun as a war it gave the character of a revolution.

Ironically, it was this revolutionary character of Tito's movement that occasioned the first clash with the Russians. As the American historian Robert Wolff, who then had access to all Western intelligence and later was able to read virtually all the written records, commented, "Though observers at the time did not doubt that Tito's bold political actions had the approval of Moscow, he had in fact acted without informing Stalin, and Stalin was very angry indeed." He wanted to put the brakes on Tito's runaway movement. At the least, he did not help it when help was needed. Thus, when the British military mission under Deakin arrived, the Partisans had been waiting for Russians. But there was no sign of Russians. As the British knew—but apparently Tito did not—when the Yugoslav royal government refused to accept them at Mihailovic's headquarters, the Russians decided not to send a mission to Tito. From his talks with Tito, Deakin thought that Tito and his staff had never understood why the Russians had not come. Then came a message ordering Tito to "Take into account [the fact] that the Soviet Union has treaty relations with the Jugoslav [*sic*] King and government and that taking an open stand against these would create new difficulties in the joint war efforts and the relations between the Soviet [Union] on the one hand and Great Britain and America on the other. Do not view the issues of your fight only from your own, national standpoint, but also from the international standpoint of the British-American-Soviet coalition."

Ultimately, the Russians had to be pushed—actually almost forced by the British—at the Tehran Conference into sending a mission to Tito. By then, Tito was on the way to victory. The Russians were of no help either to Yugoslavia's neighbor, Greece, to which I now turn.

CHAPTER 6

THE GREEK RESISTANCE

THE GREEKS HAVE BEEN KNOWN SINCE ANCIENT TIMES FOR THEIR addiction to politics. They never needed to be "politicized" because, as Aristotle commented, politics was natural to them. Combining into factions and splitting apart, acclaiming or exiling one another, constantly conspiring or actually plotting was what they did nearly full-time in each city-state. It is from the Greek word for a city-state, *polis*, that we get our word "politics." Among them, not to be passionately political was considered virtually a crime. And so deeply did the Greeks indulge in their obsession, with jockeying for status and harming rivals, that they occasionally destroyed public order. This has been true throughout Greek history but is of importance in this account in the particular ways it evolved in the aftermath of the First World War.

In 1924 a coup d'état overthrew the monarchy, which was of recent German origin, and established a republic; in the ensuing turmoil, increased by the worldwide economic depression, the republican faction was unable to consolidate its power. Seeing an opportunity, a clique of army officers intervened and restored the monarchy in 1935. As the military chiefs and civilian politicians maneuvered for advantage, the king chose one from among the generals and appointed him dictator "to save Sacred Greece from Communism."

Like the king, General John Metaxas was a strong admirer of what Hitler was then doing in Germany. Metaxas set out in July 1936, with the full approval of the king, to reshape Greece on the Nazi model. He purged the officer corps and civil service of republicans and replaced them with men ready to discipline the Greeks into what he termed the "Third Hellenic Civilization." Worried that the unruly Greeks of his generation did not provide good modeling clay to form the future Greece he envisaged, he set up a National Youth Organization (Ethniki Organosis Neolaias, or EON) on the model of the Hitlerjugend to create a new kind of Greek. And to monopolize the spread of information, he clamped a rigid censorship on the media. These efforts, he realized, would take time to reshape the nation, and he wanted to move fast. To prevent interference in his agenda, he had all identified dissidents dragged off to concentration camps. As he recorded in his diary, he believed that Hitler, Mussolini, and he were working for the same cause. But their common program was thrown off from an unexpected direction, Italy.

On October 28, 1940, without informing Hitler, Mussolini launched an attack on Greece. That attack, inspiring the Greeks with hatred of the foreign invader, did nothing to change Metaxas's domestic program, but it made the Metaxas government and the Greek people determinedly nationalist. On their own, despite the fact that a large part of the Greek officer corps, some six hundred men who were suspected of being republicans, were prevented from serving, the Greeks threw back Mussolini's botched invasion in December 1940. It was a time of great pride for little Greece, but Metaxas was alarmed that the surge of national sentiment might also be used against him, so he sought to play down the Greek victory. Even while his armies were defeating the Italians, he worked hard to assure Hitler that he would not let events play to the advantage of Britain. Presumably hoping that his German mentors would discipline his Italian tormentors, he rejected British offers to send an expeditionary force to defend Greece and made sure that the British did not use Greek facilities or airspace to attack the Rumanian oil fields on which Germany relied for much of its fuel. Greek public opinion was strongly anti-German, but Metaxas prevented any press accounts that would encourage this sentiment. Indeed, had he not died in January 1941, it seems likely that Greece would have eventually joined the Axis, as had Hungary (November 20, 1940) and Rumania (November 23, 1940), and as would Bulgaria (March 1, 1941) and Yugoslavia (March 25, 1941).

Metaxas's successors were as pro-Nazi as he had been, but the king and the new prime minister temporized out of fear that a German move against Greece seemed the clear and present danger. Their irresolution angered a group of army officers who began organizing for a coup to force the government to join the Axis; they were apprehended and their move was scotched. But, apparently, their abortive move came close enough to success to convince the government to get some form of insurance. So, in February 1941, it welcomed Churchill's foreign minister, Anthony Eden, who brought a new offer of a British expeditionary force.

The British were more worried than the Greeks about the German aggression: installed on mainland Greece and on Crete, the Luftwaffe would be able to interdict the Suez Canal. So when the Greek government halfheartedly agreed to their intervention, they moved quickly. A British force began arriving in Greece within a month. But the Greek army officer corps continued to oppose British assistance and tried to preempt the British move. The chief of staff of the northern Greek army passed word to the Germans that if they would rein in the Italians and stabilize the Albanian frontier, the Greek army would expel the British. The officer corps was not alone: the prime minister secretly made a similar offer to Hitler's ambassador. All this sub-rosa espionage is important for my purposes here in showing that the king, the civil government, and the army and the senior ministers, while hostile to Italy on nationalist grounds, were willing either to join the Axis or at least to enter into friendly relations with it.

Attempts to reach an understanding with Germany were defeated by the actions of yet another outside player: on March 25, Yugoslavia's royal government joined the Axis. The next day, the Yugoslav officer corps made an anti-German coup d'état; that was perhaps the first truly popular action it had ever taken. The outburst of joy among civilians, although favorable to the generals, alarmed them because they feared "the people" at least as much as they disliked the Germans. So they disavowed their own action by assuring Hitler that they would abide by the previous government's decision to join the Axis. It was too late. Hitler was furious. Worse, as I mentioned in the previous chapter, he was embarrassed because he was then entertaining the foreign minister of his new ally, Japan. He was even more disturbed when the Yugoslav government, also seeking insurance, signed a pact of friendship with Stalin.

So, although his general staff advised against it, Hitler decided to attack Yugoslavia as the first stage of a move all the way south and east toward the Middle Eastern oil fields. In twelve days the Wehrmacht blitzkrieg had cut through Yugoslavia and reached the Greek frontier on April 6, 1941. In seventeen days the Greek army collapsed and the British expeditionary force got out as best it could, losing twelve thousand soldiers and all their equipment. It was a stunning victory, but despite the best efforts of Metaxas and his successors, Hitler and Mussolini had made "Greece" a part of the "Free World."

As the Germans raced in, the king, some of Metaxas's associates, part of the general staff, and what was left of the army fled the country to the sanctuary of British-controlled Egypt. With them went "Greece." In the eyes of Britain (and also America, which followed the British lead), "Greece" was no longer physically in Europe but had moved, in the person of the king, to a British safe haven. The geographical Greece, then under German occupation, was an as-yet-undefined territory, waiting to be resurrected when the British drove out the Germans.

What was happening to the seven and a half million people left behind inside Greece was not known to the Allies. But events there were seminal. As information filtered out, it became clear that the Germans had installed a puppet government composed of Metaxas's remaining ministers, civil servants, and police. Overcoming the shock of defeat, there were many Greeks who were eager to work with the Germans. The Greek Fascist Party was thought to have then had about fifty thousand members, and there were various extreme right-wing terrorist groups like the X (pronounced in Greek "khi") which were more than willing to use the opportunity afforded by the German occupation to defeat or kill their old political enemies. A collaborationist government, first under the general who had surrendered to the Germans and then under Metaxas's deputy, Ioannis Rallis, was formed on April 21, 1941, to administer conquered Greece.

Administering Greece under Nazi orders translated into starvation for thousands of Greeks because the Germans requisitioned all available foodstuffs. Unemployment soared as factories were dismantled and their machinery shipped to Germany. And Greece's merchant marine hauled up the German flag. As Mussolini commented to his foreign minister, "The Germans have taken from the Greeks even their shoelaces . . ."

But while humbled, national pride remained fierce. The Germans, naïvely, like most invading armies, thought that they would be welcomed with smiles and flowers; they were apparently shocked by what happened. Within days, the first act of sabotage, the explosion of two ships carrying ammunition, was carried out in the port of Piraeus, and shortly thereafter, in a stunning symbolic act, two young Greek students tore down the Nazi flag from the Acropolis. These brave acts electrified Greek nationalist sentiment, but resistance remained symbolic and private. There was no movement and no leaders. The government-in-exile was out of touch and was hardly an exemplar since it had tried to work with the Nazis and, under Metaxas, had done what it could to destroy Greek liberty. Within occupied Greece, Metaxas's heirs were outright collaborators, employing the political police and gendarmerie against anti-Nazi Greeks at the behest of the Germans.

Little is known of them even now, but small groups of men, hating the Italians, fearing the Germans, or outraged by the collaboration of other Greeks, took to the mountains. At first they were voiceless and isolated from one another. Then, in September 1941, the first call for a national resistance movement was sounded. At the urging of the few members of the tiny Greek Communist Party (Kommounistikon Komma Ellados, or KKE) who were still at liberty, but in response to widespread nationalist ardor, an underground movement known as National Liberation Front (Ethnikon Apeleftherotikon Metopon, or EAM) was formed in February 1942.

Like all insurgent movements, EAM was initially minuscule—at conception it was said to have numbered only fifteen men. These men had learned their political tradecraft in the underground, fighting Metaxas's dictatorship. They were mainly urban and were more accustomed to issuing statements than fighting. But they realized that they could not fight the Germans just with leaflets. We do not know how they hit on their new form of resistance, but on October 11, 1942, they took the initiative in the little village of Kleitsos, forming a village-based organization that became the prototype for the underground. The committee, made up of residents, focused the energies of the villagers on concrete, local problems and thereby laid the foundation for what came to be called "Free Greece."

Each village was expected to elect a committee headed by a "responsible person" (Greek: *Ipefthinos*), a local man who was almost invariably a member of the KKE. The responsible persons from each village in a group of villages then got together and elected one of themselves as the *Ipefthinos* for their group; he then met with his counterparts in neighboring villages and elected an *Ipefthinos* for that area and so on, until the provinces elected twenty-five delegates to the central committee of EAM. All Greece, or as much of it as came under control of the resistance, was thus formed into a political pyramid. Ultimately, before the end of the German occupation, it is believed that EAM was composed of virtually the entire adult population—somewhere between half a million (the British intelligence estimate) and two million (the EAM claim) in a population of seven and a half million. EAM had become, both by government default and by political action, *the* national institution.

After the civil war, it was customary to describe EAM as a Communist party. There is some truth to that description: it was certainly founded by the KKE, and the KKE used the pyramid structure and the role of the "responsible persons" to guide its policies, but German intelligence, which was constantly engaged in finding out everything about the resistance, estimated that only one in ten members and about one in four leaders of EAM was a Communist. In form at least, EAM had become a coalition with members from four of the political parties that had existed under the Metaxas dictatorship. The Greek Church was represented by six bishops and a large number of priests, along with many of Greece's most respected academic and business leaders. In these statistics appears a contrast to Tito's Yugoslavia, where all power was in the hands of the Communists.

The military wing of EAM was known as the National Popular Liberation Army (Ethnikos Laikos Apeleftherotikos Stratos, or ELAS). A surprising number of its members were senior officers of the former regime's army. The Greek-American historian L. S. Stavrianos has identified among them sixteen generals, thirty-four colonels, and fifteen hundred other commissioned officers. ELAS was controlled by a three-man committee. The military member was a former colonel in the Greek army (who had been imprisoned by Metaxas); he was joined by a *kapetanios* (political or administrative leader) who dealt with relations with civilians, and by a representative of the EAM central committee. Both of the latter two men were members of the KKE. This structure was more or less duplicated, depending on local conditions and available personnel, throughout the structure of ELAS. Given the problems of communications, which were severe, the structure was highly centralized and firmly under control of EAM and, through it, of the KKE. The analogy with Tito's organization in neighboring Yugoslavia is striking, but it is equally striking that Greece lacked a Tito: the secretary-general of the KKE, Nikos Zakhariadis, who might have played Tito's role in the resistance, was in the German concentration camp Dachau until the end of the war. It was in the foot soldiers, not the generals, that the two movements were comparable.

Most of the *andartes* (combatants) were poorer people, mainly farmers, drawn from rural areas. Estimates vary as to the number of actual combatants in Greece. The British thought ELAS eventually numbered about fifty thousand, while EAM claimed eighty-five thousand. German intelligence estimated about twenty thousand men and women backed up by a larger force of "reservists." As in other guerrilla wars, people drifted in and out of the resistance, going home to sow or harvest crops, fighting or lying low as German military activity waxed or waned, so exact numbers are not known and would, in fact, be almost meaningless. But the order of magnitude is impressive. In Yugoslavia and Afghanistan, the numbers are comparable; in Algeria and Northern Ireland, the numbers of combatants were far smaller; in Vietnam and of course in China, they were much larger. What was important was that ELAS came to control virtually the entire country except for the main cities.

The village committee, as in Yugoslavia, was the foundation of the whole structure. It took charge of collecting taxes either in money or in kind to support the combatants. To keep the committee honest and to make its work transparent, it was required to post an accounting in the village square or prominent place showing how much tax each villager paid. Then, as the resistance spread and grew in intensity, committees were expected to divide what they had collected into three parts: roughly a fifth was set aside for administration of the village, a small fund was established to aid villages that had been ravaged by the German army, and three-quarters went to the military organization, ELAS, to enable its members to fight rather than to farm.

While EAM was organizing itself and growing, what remained of the Metaxas bureaucracy, gendarmerie, political police (who were patterned on the Gestapo), and paramilitary organizations (patterned on the SS) collaborated with the German occupying forces, helping them to hunt down EAM/ELAS figures, to quell, jail, or kill those involved in unrest or suspected of being in contact with the British. There is no evidence that these domestic heirs of Metaxas were in touch with the royalist government-in-exile, which the British had set up in Cairo, but later events showed that they remained loyal to the royalist/Metaxas regime as it existed before the German invasion.

Also opposing EAM/ELAS inside Greece was a rival group, the National Republican Greek League (Ethnikos Demokratikos Ellinkos Syndesmos, or EDES), headed by Napoleon Zervas, which, despite the name and Zervas's past affiliation, was a royalist organization. It was comparable to the Yugoslav Cetniks under Drazha Mihailovic. Like the Cetniks, EDES was the recipient of British arms and money, and also like the Cetniks, it collaborated with the German occupiers. So unwilling was Zervas to enter the war against the Germans that he did so in June 1942 only after the British bribed him with what was then a fortune in gold and his British "handler" (or as the CIA called such a man, his "case officer"), Colonel Montgomery Woodhouse, threatened to denounce him to the Gestapo if he did not. Then, having organized his movement with British arms and money, he saw it nearly disappear through casualties and defections by December 1943. Woodhouse then proposed to his superiors in Cairo that it be disbanded. They rejected his advice, and he switched to become its strongest British supporter.

Despite Zervas's spotty record and EDES's reluctance to fight, it was EDES into which the British funneled most of the aid they delivered during the occupation. Then, although funded and armed by the British, it reached an understanding with the Germans to limit their attacks. In relative security and wealth, it expanded so that by 1944 it was thought to contain about seven thousand active guerrillas. The difference in the British treatment of the Cetniks and EDES was of fundamental importance to events in Yugoslavia and Greece, but of equal importance were the differences of the British relationships with Tito's National Liberation Movement and the Greek EAM/ELAS. A close examination of these relationships will illuminate the nature of guerrilla warfare.

Roughly the same size as Tito's Partisans, EAM was either the first or second largest resistance movement in Europe, dwarfing, for example, the contemporary French resistance, which numbered only a few thousand combatants until after the Anglo-American invasion and never controlled any significant territory. However, throughout the war, even after they had decided to support Tito's completely Communist government, the British remained hostile to the EAM. It has often been said that this was because of Churchill's strong royalist predisposition. His feelings may have played a role in his decisions, but there was an objective reason for the British policy. Since it was so significant in what happened in Greece during the German occupation, the British invasion, and the subsequent American counterinsurgency, it needs to be understood to evaluate the resistance.

British policy in the Mediterranean and the Middle East became fixed during the time of Napoleon. When he invaded Egypt at the end of the eighteenth century, the British feared that he was on the road toward Britain's Indian empire. That threat became fixed in British strategy so that even after Napoleon had been defeated, Britain remained determined to stop any power from pushing south from Europe. First came the threat of the Russians, then in the Second World War from the Germans and, foreseeing the cold war (as Churchill already did even before the war), from the Soviet Union. India made the Suez Canal a major British concern. By the First World War, when the Royal Navy converted from coal to oil, the Iranian oil fields also became "vital" British interests. They then joined India as the assumed targets of German and Russian aggression. To counter the danger, generation after generation of English soldiers and statesmen engaged in what became known as the "Great Game," the cold war in Afghanistan and other parts of Central Asia, propped up the dilapidated Ottoman Empire, and built a second line of defense in the puppet states of Iraq, Kuwait, and Egypt. In 1941 the Germans had come close to breaching all these positions. So, as the war evolved, Churchill had come to look upon Greece as the only possible "stopper in the Mediterranean bottle."

Would the stopper hold? To be sure that it would, Churchill decided that he must control Greece. He realized that Britain had a much better chance to control a royalist government (as Britain *was then doing* in Iraq, Jordan, and Egypt) than it would to control a mass national movement (as it *was then failing to do* in India). EAM looked like it would be at least as difficult to control as Gandhi's Indian independence movement. Indeed, as it acquired power, it proclaimed its intent to remain independent, which Churchill interpreted to mean that under its control Greece would establish a close relationship with the Soviet Union.

Churchill received a warning on the danger of an independent Greece from one of the few men he admired, South African prime minister General Jan Christiaan Smuts. General Smuts had a particularly close relationship with Queen Frederika of Greece and had appointed himself Churchill's adviser on Greek affairs. In a striking harbinger of the later "domino theory" that would haunt American officials on Vietnam, he warned Churchill that "with politics let loose among those people we may have a wave of disorder and wholesale communism set going all over those parts of Europe. This may even be the danger in Italy, but certainly in Greece and the Balkans." Churchill immediately sent a message to the Greek king, saying, "We are all looking forward to your returning to Greece at the head of your armies and remaining until the will of the Greek people is expressed under conditions of tranquility." His policy was soon challenged by the EAM, which, on March 10, 1944, announced that it had established a provisional government.

The establishment of a provisional government, known as the Political Committee of National Liberation (Politiki Epitropi Ethnikis Apeleftherosis, PEEA), inside Greece and unaffiliated with the Metaxas legacy, electrified the Greek armed forces then in Egypt. They immediately petitioned to serve as its army rather than serving as the army of the royalist government-in-exile. When the British refused, the army-in-exile mutinied. In riposte, the British disarmed them, hanged (at Churchill's personal order) many of their officers, and put others in prison. Seeing EAM as the principal enemy of the defense of the British Empire in the Middle East, Churchill referred to its members and their supporters as "miserable Greek banditti" and "treacherous, filthy beasts." But he was more than angry at their treachery; he was frightened by their move toward self-determination. The British stopper was in danger of popping out of the Greek bottle.

The way to secure the stopper was evident: the Allies must mount their counteroffensive to retake Europe through Greece rather than through Italy or France. That way, they could anchor in Greece a defensive barrier—not unlike the Soviet organization that Churchill later called the Iron Curtain—stretching across Turkey and Iran to the northern frontier of India to block the Russians from "British Asia." But the Americans, on military grounds, opposed an attack on the Germans through Greece; they favored assaults first on Italy and then on southern France. Churchill argued vigorously against Eisenhower and the other generals but lost the argument. Thwarted in his first plan, Churchill conceived an alternative: a grand deal with the Soviet Union reminiscent of nineteenth-century diplomacy. He would recognize that Stalin had a strategic imperative to protect the western frontier of the Soviet Union in return for Russian acknowledgment of the British strategic imperative to protect the northern flank of its empire. So crucial was this conception to Greek insurgency that it deserves to be recounted here.

When the Americans refused to be drawn into his eastern strategy and went ahead with the commitment of Allied troops to Western Europe, Churchill instructed Foreign Minister Anthony Eden to propose on May 5, 1944, to Soviet Foreign Minister Vyacheslav Molotov a division of spheres of control: Britain would agree to the Soviet Union "temporarily" taking charge of Rumania if it agreed to Britain taking charge of Greece. Following up, Churchill met Stalin at the Kremlin on October 9, 1944; there the two leaders formally agreed on one of the most sweeping deals known in history. Although the deal was made in 1944, there are signs that it had been hinted at before.

The Russians were cautious. They wanted to be sure that the United States also approved. It did not. Secretary of State Cordell Hull was strongly opposed. Ultimately, however, at Churchill's urging, Roosevelt overruled Hull. With Greece secured, Churchill told his personal emissary to Tito, Brigadier Fitzroy MacLean, that he was prepared to "write off" Yugoslavia. As MacLean recounts, when he reported to Churchill that "Tito and the other leaders of the

Movement were openly and avowedly Communist and that the system which they would establish would inevitably be on Soviet lines and, in all probability, strongly oriented toward the Soviet Union," Churchill replied with a flippant remark but one that shows how much he was focused on the strategic role of Greece, a role that Yugoslavia could not play. Since it could not, what happened to it was of less importance than what happened in Greece to the British Empire and its Asian assets. As MacLean recalled:

> "Do you intend," he asked, "to make Jugoslavia [*sic*] your home after the war?"
>
> "No, sir," I replied.
>
> "Neither do I," he said. "And, that being so, the less you and I worry about the form of government they set up, the better. That is for them to decide. What interests us is, which of them [Tito's Partisans or Mihailovic's Cetniks] is doing most harm to the Germans?"

It was this view of imperial strategy that was to govern Churchill's approach to the Soviet leaders. His view was a classic example of realpolitik, brutal in its assessment of the reality of "boots on the ground," grand in its sweep, and stunning in its audacity. His account of the meeting he and Foreign Secretary Eden held, without any staff members, with Stalin and Molotov is a remarkable document. In it he wrote that he said to Stalin:

> "Let us settle about our affairs in the Balkans . . . Don't let us get at cross-purposes in small ways. So far as Britain and Russia are concerned, how would it do for you to have ninety per cent predominance in Rumania, for us to have ninety per cent predominance in Greece, and go fifty-fifty about Yugoslavia?" While this was being translated I wrote out on a half-sheet of paper [these numbers and a similar arrangement for Hungary and Bulgaria] . . . I pushed this across to Stalin, who had by then heard the translation. There was a slight pause. Then he took his blue

pencil and made a large tick upon it, and passed it back to us. It was all settled in no more time than it takes to set it down.

Although he may not have known of the deal at the time, the future prime minister Harold Macmillan, then the resident minister in the Middle East, lamented, "All the rest of Balkans are gone. This [control of Greece] is our last chance of avoiding the establishment of a Communist society." From then on Churchill could absolutely not afford to lose Greece. EAM had to be destroyed. He wrote Anthony Eden on November 7, 1944, "In my opinion, having paid the price we have to Russia for freedom of action in Greece, we should not hesitate to use British troops to support the Royal Hellenic Government . . . I fully expect a clash with EAM, and we should not shrink from it, provided the ground is well chosen."

Neither the British nor the Russians told the Greek insurgents of the Churchill-Stalin deal. Stalin did, however, say to Milovan Djilas, Tito's emissary in Moscow, "The uprising in Greece has to fold up . . . The uprising must be stopped, and as quickly as possible . . . we should not hesitate, but let us put an end to the Greek uprising." Having made a profitable exchange, Stalin kept his side of the bargain, as Churchill later confirmed. Belatedly it appears, for documents (if there were any) are not available, the Russians sent a message to the EAM to inform its leaders that unless they reached a deal with the British, they would be crushed and Russia would do nothing to help them.

Meanwhile, the Germans began to pull back from Greece. In their place, the British rushed in what remained of their forces in Egypt and those elements of the Greek army they regarded as reliable. Within weeks, there were five thousand British troops in Athens, and on December 5 Churchill instructed their commanding general: "You are responsible for maintaining order in Athens and for neutralizing or destroying all E.A.M.-E.L.A.S. bands approaching the city . . . *Do not however hesitate to act as if you were in a conquered city where a local rebellion is in progress.*"

Isolated, feeling betrayed and exhausted by the underground

war, the EAM caved in and agreed to place ELAS under the royalist government-in-exile. The British immediately began to disarm ELAS while leaving armed the German-organized and formerly German-commanded "Security Battalions," which had rounded up and shot Greek hostages during the occupation. These formations now had ELAS literally in their gun sights.

Push came to shove in the central square of Athens on December 3, 1944. The police, who also had worked under Nazi control during the occupation, fired into crowds of demonstrators. ELAS tried to regroup to resist, but carefully avoided conflict with British troops. A last-minute compromise was worked out for a cease-fire: ELAS would disarm in return for a program that included a plebiscite on the monarchy, promise to prosecute the collaborators, and guarantees on upholding of civil rights. The compromise quickly turned out to be a hoax as vigilante bands from the "National Guard" and the X began hunting down and killing the disarmed ELAS veterans.

Two questions arise at this point: since they had supported EAM/ELAS against the Germans through the terrible years of the occupation, why did the Greeks not more strongly support them in 1945, and why did the leaders of EAM/ELAS give in?

The first question can be answered only speculatively. I believe the reason, however, can be seen in all guerrilla movements. Once the hated Germans and the despised Italians pulled out, the main reason the Greek population had supported EAM faded. Most Greeks looked upon the British as liberators rather than as occupiers, so EAM could not mobilize support against them to the same degree as it had against the Germans and Italians. And the population was tired, hungry, and desperate for peace. It no longer was prepared to sacrifice for EAM/ELAS.

The second question, why the insurgent leaderships foundered, conjures a more interesting answer. At the end of the war, the central committee of EAM/ELAS followed the Russian line, which favored a territorial concession in northeastern Greece to the by-then Communist Bulgaria. This policy aroused almost as much anger as had the German policy of giving northwestern Greece to Fascist Italy.

Over this nationalist issue, the KKE split. The KKE Secretary-General Nikos Zakhariadis, newly released from Dachau concentration camp, returned to Greece to impose the Soviet line. That line was rejected by ELAS's military leader Markos Vafiadis. As Tito was then beginning to do in Yugoslavia, Markos staked out a new form of "national communism." On December 24, 1947, Markos made the split permanent by announcing the formation of a "Provisional Democratic Government."

Meanwhile, hammered by the British and the Greek army and police loyal to the Royalist regime, group after group of non-Communists fell away from the wartime coalition until virtually all that was left of that EAM/ELAS coalition was the bitterly divided Communist core. As its power faded, the remnant of ELAS was pushed out of the cities into the rural areas; there, the peasantry was politically and religiously conservative. It was motivated by no great cause like land distribution, which energized contemporary Filipino and Vietnamese peasants. Viewing the insurgents and the government with nearly equal mistrust and fear, it just wanted to be left alone, as I discovered firsthand.

When I first visited Greece in May 1947, I went down to the area said to be dominated by ELAS, the Peloponnesus. I traveled all over the area with a mobile hospital team and had a chance to speak with dozens of villagers. Almost to a man, they expressed hostility to ELAS. A typical remark was "we got along with the Germans. Italians were much harder. The *andartes* killed about three thousand people down here including everyone who collaborated." Everyone with whom I talked professed to be a royalist, and they were certainly deeply religious. They were also obviously exhausted by the war, hungry, ill-clothed, and ready to submit to anyone offering an end to their suffering. Once the Germans had left, the fire had gone out of the nationalist movement.

But ELAS struggled on. Using what American military advisers called "evasive" tactics, that is, guerrilla warfare, the relatively small remaining guerrilla force—estimated at 15,000–18,500 men and women in 1948—was wearing down and demoralizing the Royalist army. The Americans observed that, although it had a ten-to-one superiority in men and matériel, the Royalist army lacked a "will to fight." It was in this situation in March 1948 that the Americans took up Churchill's strategy, clothing it in suitably American phrases as the Truman Doctrine, to help "free peoples resist subversion by armed minorities." Under the Truman Doctrine, America became the principal financial and military supporter of the Royalist government. The Greek civil war was to be America's first counterinsurgency war since the Philippines half a century earlier and a model for its fight in Vietnam twenty years later.

Americans were quick to take credit for the suppression of the insurgents. The head of the American mission, General James A. Van Fleet, sounded that claim in an article entitled "How We Won in Greece." The lesson that was later to be applied to Vietnam was spelled out in a series of five articles entitled "The Anti-Bandit War" by Colonel J. C. Murray, USMC, in *The Marine Corps Gazette* beginning in January 1954. Van Fleet's and Murray's analysis was virtually all military: American "advisers" whipped the Greek army into shape; provided it with large amounts of equipment, including artillery, tanks, and supply vehicles; poured in money to win over officials; and provided Greek forces with military aircraft flown by American pilots. American advisers convinced the Greek military that they had to "clear and hold" if they were to win. That same catchphrase is one we now hear for Iraq.

Emphasis was given to closing the frontiers to starve the guerrillas and to deny them a safe haven in which they could rest and regroup. On their own, the Greeks were unable to accomplish this, but after his split with the Russians in 1948, Tito became less willing to help EAM/ELAS. To the limited degree he actually did close the frontier in July 1949, this was undoubtedly a factor in the decline of guerrilla effectiveness, but it was not, as American writers have so often said, decisive. Most guerrilla movements have been isolated, but rarely has isolation proved fatal. Outside aid makes life easier for the guerrillas but is not decisive. What was fatal to EAM/ELAS was political. What amounted to an act of suicide was forced upon

EAM by the Soviet Union: EAM was put in the position of opposing Greek nationalism when in March 1949 it espoused the Cominform-supported "People's Liberation Front of Slav Macedonia." This act, which violated Greek nationalism—from which, rather than from Communism, its popular support had derived—split EAM.

In an attempt to overcome the split in its ranks, EAM purged the commander of ELAS, Markos Vafiadis, who even with reduced means had been fighting a successful guerrilla war. In replacing him, the KKE leadership decided to fight a conventional campaign. The change of tactics was militarily disastrous. As the American observer, Colonel Murray, wrote, "The decisive defeat of the guerrillas was made possible by their departure from proper guerrilla organization and tactics in their effort to defend the base areas along the northern frontier and the gathering of their one-time small bands into larger formations. . . ." When they tried to stand and fight, they lost the advantage of the guerrilla. Murray was right: the change of tactics was important, but crucial was the loss of popular support and by the split within EAM brought about by the policy the Soviet Union imposed upon EAM. Neither of these could have been engineered by the Greek government or their American advisers. As a result of these unintended gifts of the Russians, the back of the insurgency was broken by the end of August 1949.

Both the Americans and the Greek government were elated. They had found a formula to stop Communism.

Or had they? When they tried to apply the Greek formula to Vietnam ten years later, it did not work. It was not the closure of the frontier, massive aid programs, or even engagement of ground troops that was crucial. The one thing that the Americans were unable to do in Vietnam was what had really counted in Greece—splitting the resistance. Ironically, the American victory in Greece was the gift of the Soviet Communist Party.

CHAPTER 7

KENYA AND THE MAU MAU

THE ORIGINAL BRITISH INTEREST IN EAST AFRICA WAS SIMILAR TO THE British objective in Greece, to protect the route to India. Britain was disturbed by the French acquisition of Madagascar and thought it needed a base on the Indian Ocean. The British statesmen thought that Mombasa would be a convenient anchor for the Royal Navy, but the British admirals did not think that port worth the cost of taking it. Since cost was the determinant, the British welcomed a way to get it at a bargain price. Ironically, Germany offered the means. The German chancellor Prince Otto von Bismarck wanted a slice of East Africa, and his move enabled the British to piggyback on German imperialism. The two powers convinced the Sultan of Zanzibar to cede his rather vaguely held African empire to them: the southern part (renamed Tanganyika) went to Germany and the northern part went to the British.

At that time, the north was inhabited by several groups of Africans. The largest, known in the West as Kikuyu (who called themselves the Gekoyo), were speakers of a dialect of the Bantu language family and were mainly agriculturalists. Their principal occupations were growing millet and raising cattle. They were not villagers but lived in *mbari,* close kinship groups in clusters of dwellings. Beyond these clusters, the Kikuyu felt an affinity to two other groups: "vertically," they ascribed to themselves tribal identities, and "horizontally," each man, upon coming of age, entered a sort of fraternity composed of age-mates. Overall, however, in no effective sense were the Kikuyu a nation. This was true even when the Kikuyu were defending themselves against their local enemies, the aggressive, nomadic Masai. Nor does there seem to have been any sort of association with their other Bantu-speaking neighbors. This fracturing of the population into kinship and tribal groups was to play a key role in European settlement and subsequently in resistance to it.

In the 1890s, two major shifts occurred in the Kikuyu area. First, there was an outbreak of epidemics. As among American Indians after the arrival of Europeans, smallpox took a devastating toll, and the ensuing famine weakened the survivors. Consequently, the Kikuyu moved out of large parts of the area they had been farming. It was this area, later known as the White Highlands, into which the British began to encourage Englishmen, Eastern European Jews, and South African Boers to settle. This influx of whites was the second major shift in Kenyan society. It was resisted by the Kikuyu as they had resisted raids made by the Masai. But as viewed by the British, resistance was an assault on public order and must be suppressed. Both the resistance and the suppression were only sporadic and limited. The East Africa Protectorate, as it was then known, did not seem to the government in London sufficiently important or attractive to merit much attention. But after 1902 when the fertile uplands around Lake Victoria were added to the East Africa Protectorate and a new railway from Mombasa was opened in 1903, Kenya began to acquire importance. Thereafter, European settlement surged.

The incoming Europeans did not wish, personally, to work the land they began to acquire; they wanted to work their new plantations with cheap local labor. In this, they resembled the colonists who had come to North America, and no more than the American Indians did the Kikuyu wish to work for the new arrivals. So the British issued a sequence of edicts to compel them to do so. One aspect of this general policy, again resembling colonial America, was to remove the Kikuyu from lands they regarded as theirs to what Americans called "reservations" and the British termed "native reserves."

Race made such policies seem appropriate to the white community. The natives were blacks. The white settlers simply did not regard them as equivalent human beings: the natives had no inherent rights to the land and, in any case, they did not use it efficiently. The wonderful resources of Kenya were wasted on them, whereas whites would make them productive. The Kikuyu seemed relatively pliable, even pitiable. Among the natives, only the Masai commanded grudging respect, but since no white wanted what the Masai had, semidesert lands suitable only for raising cattle, they were irrelevant to the white-Kikuyu conflict. At first, even that conflict was low-key since, being relatively few, spread over a large area, and unsure of the support of the government in London, the settlers tread carefully. But by the end of the First World War, the white community reached a politically critical mass. Recognizing its growing importance, wealth, and influence, the British government in 1920 established the area as a colony, named after the great mountain, Kenya, for the benefit of European settlers. The natives could not be got rid of, but they could be put to work as virtually another form of domestic animal or pushed aside to the native reserves.

Haltingly, as the small group of Kikuyu who had been educated by missionaries became conscious that they were being ruled by and for the European settlers—with the natives prohibited from growing the main cash crop, coffee; unable to buy land in areas farmed by their ancestors; artificially but legally held to low wages; allowed to travel only with settler-controlled passbooks; and subjected to other discriminatory practices—they formed the first of several organizations that grew into the Kikuyu Central Association.

The political and even the intellectual evolution of the Kikuyu was, of course, opposed at every step by the white community. The Kikuyu's first clash with the colonial authority came in 1920 when a crowd, peacefully protesting the arrest and exile of one of its leaders, was fired upon by uniformed police and a group of white settlers. In part focused by this incident, the whites' view of the blacks hardened further and the Kikuyu grew more fearful. They had reason to be more fearful. In a ruling that swept away most of their surviving economic and legal defenses, the colonial High Court ruled in 1925 that the 150,000 Kikuyu "squatters" had no rights under the traditional form of ownership (known as *githaka*) even of *shambas* ("kitchen gardens") in areas taken over by whites. That ruling was followed by the report of the government Land Commission of 1934 that affirmed European title to virtually all the good agricultural land in the colony. The full impact of these two rulings was not realized at first but was increasingly felt as the growth of the Kikuyu population began to cause severe overcrowding on the remaining lands.

At the outbreak of the Second World War, Kenya's blacks numbered about five million and the whites fewer than thirty thousand. The administrative structure—there was no real political structure because Kenya was ruled by a British appointee—was the inverse of the population: the whites completely controlled the colonial militia, the police, and the judiciary, and served as advisers to the British governors. The blacks were totally excluded except as foot soldiers in the militia and as chiefs, appointed by the British to enforce decisions on their tribesmen.

It was in this context that the government decided forcibly to evict more than one hundred thousand Kikuyu from the white area. As all the suitable agricultural lands were taken, there was no place for them to go but to the already overcrowded Kikuyu reserve or to shantytowns around the main cities. As the natives came to understand the effects of these rulings, an explosive situation was created. Hearing rumblings of discontent and replying to settler appeals, the British banned the Kikuyu Central Association as a wartime "security measure" in 1940. Thus the natives had lost their only peaceful means of expressing grievances.

Once the Second World War ended, the justification for preventing black association lapsed, and the British allowed a new organization to form. This one aspired to be a real political party known as the Kenya African Union. The beginnings of this organization were unpromising. Only a dozen men were the prime movers. All were Kikuyu and, apparently, all were moderate men, of whom the best known was Jomo Kenyatta. Kenyatta would subsequently play a major role in Kenyan affairs, but at that time he had been for seventeen years in exile in London and so was out of touch with his fellow countrymen. Nonetheless, he was selected by the small committee of the Kenya African Union to become its president.

With changes of names and dates, this sequence of events might be read as a generic history of resistance movements in British, French, Belgian, and Portuguese Africa. As in other areas of Africa, there were scattered acts of violence, but for the most part these amounted only to personal vengeance. What happened then, however, was distinctively Kikuyu. Led by the younger graduates of the mission schools, increasing numbers of Kikuyu began to search for some way to organize themselves. Unlike resistance movements in most other areas of Africa and Asia, they found no stimulus or support in nationalism, religion, or Communism. What they did have was a traditional form of association with age-mates that was both symbolized and effected by the ceremony of oath taking. Although, traditionally, such "oathing," as the British called it, while *linking* those of the same age, *divided* those not of the same age. But it lent itself to a wider use. As it came to be modified, it gave rise to an informal but real sense of nationhood. As in all insurgencies, Kenya began modestly. In 1950 a group of only a dozen of the younger activists, who called themselves the Kiambaa Parliament, took the first steps beyond the Kenya African Union toward the creation of a resistance organization.

In his excellent study of the Kikuyu resistance, which came to be known as the Mau Mau, David Anderson wrote that "the very words [Mau Mau] conjure up memories of something evil lurking in history's dark shadows. Mau Mau was the great horror story of Britain's empire in the 1950s. The battle to suppress the revolt in Kenya was presented as a war between savagery and civilization . . ." Echoing the attitude of the English settler community, a mixture of fear and loathing, the writer Elspeth Huxley called Mau Mau "the yell from the swamp." The war between the natives, the whites, and the British authorities that engulfed Kenya from 1952 to 1960 was certainly dirty and brutal, but the facts belie the lurid accounts of the time: in the seven years it lasted, only thirty-two European settlers and fewer than two hundred British soldiers and police were killed. No casualties, of course, are trivial, but why did these evoke such blood-chilling rhetoric? The main reason, I think, is that some of the insurgents were former household servants of the white settlers, who had believed them to be backward and childish, but also loving and kindly, and above all loyal to their white masters. "It was," wrote Graham Greene, "as though Jeeves had taken to the jungle." In the eyes of the whites, that these apparently faithful retainers should revolt against their benevolent employers was the ultimate treachery. It was the proof of the racial stereotype the settlers had firmly in mind.

A second reason that the revolt evoked such fury from the white settlers is that, to try to overcome the lack of coherence in their movement and the lack of a sense of nationalism or commitment to an ideology or religion that, as we have seen, gave unity to other insurgencies and resistance movements, the Mau Mau leaders tried to conjure unity by means of what whites saw as black magic. While they would not have used the same words, the blacks implicitly agreed. The oathing ceremony aimed to so traduce normal codes of behavior as to psychologically mark the oath taker. In participating, the oath taker moved across the mental barrier that had constricted his actions and figuratively emerged as a different person. Newcomers were forced to perform acts, sometimes revolting or obscene acts, that welded them into the rebel brotherhood. In a structural sense, oathing transformed the participant just as engaging in terrorism or individual murder, which similarly violated social and cultural norms, did in other insurgencies: militants were altered by the act and joined with others who were similarly changed into a new association from which it was supremely difficult, perhaps even self-destructive, to withdraw. Thus, the forms of oath taking, while fascinating to anthropologists, horrified those who did not participate.

One of the men who tried to understand the nature of the

oath-taking ritual was Josiah Mwangi Kariuki, who participated in Ndemwa Ithatu, the ceremony of the oath of unity. As he describes it, the words of the oath and the ritual surrounding them essentially accomplished a *political* objective: to create group cohesion. They made the oath taker swear (upon threat of magical retribution) to fight the enemy and protect fellow combatants. Kariuki recounts that he was led by a friend into an isolated clearing in the forest. There, after being "hit about a little" by those who had already taken the oath, he and the other newcomers "passed through the arch [formed by banana tree branches] seven times in single file" and then "bowed toward the ground" as the oath administrator "circled our heads seven times with the meat [of a dead goat], counting aloud in Kikuyu. He then gave each of us in turn the lungs [of the goat] and told us to bite them. Next he ordered us to repeat slowly after him the following sentence:

> I speak the truth and vow before God
> And before this movement.
> The movement of Unity,
> The Unity which is put to the test
> The Unity that is mocked with the name of "Mau Mau,"
> That I shall go forward to fight for the land,
> The lands of Kirinyaga that we cultivated,
> The lands which were taken by the Europeans
> And if I fail to do this
> May this oath kill me,
> May this seven kill me,
> May this meat kill me.
> I speak the truth that I shall be working together
> With the forces of the movement of Unity
> And I shall help it with any contribution for which I am asked,
> I am going to pay sixty-two shilling and fifty cents and a ram for the
>   movement
> If I do not have them now I shall pay in the future
> And if I fail to do this

May this oath kill me,
May this seven kill me,
May this meat kill me.

The oath takers were then anointed on the forehead with blood as the oath administrator explained "that he did this to remind us that we were now fighting for our land and to warn us never to think of selling our country." Next the oath administrator made three tiny scratches on the wrists of the men and put pieces of meat on them so that a few drops of blood got on the meat. "Next he gave us in turn the meat to bite and while we were doing this he said, 'The act of eating this meat with the blood of each one of you on it shows that you are now united one to the other and with us.' This was the end of the ceremony . . ."

David Anderson recounts testimony that in place of the dead goat, the Mau Mau at least once were reported to have tried to terrify their opponents by using the body of a man they had murdered for refusing the oath. Anderson says that although this was an "utterly exceptional case," it was "added to the tales that circulated among white settlers of the perversion and horror of Mau Mau oathing." To get at an understanding of the objective of the Mau Mau, we must dig under the then prevalent image of savagery, treachery, and viciousness. Certainly the Mau Mau had objective reasons to revolt. The European and South African settlers had taken lands that the 1.5 million Kikuyu regarded as their national patrimony and had converted virtually the entire Kikuyu people into serfs to work it for them. The Kikuyu had no civil rights and were subjected daily to arbitrary police and settler humiliation and violence. They had no effective say in the management of their own affairs and certainly none in managing Kenya. Moreover, while at that time other areas of Africa seemed to be moving toward freedom—*uhuru*—Kenya looked to be poised to slip even further under white control as was then happening in South Africa and Southern Rhodesia. Even where Britain loosened the reins of its empire, it was not natives who benefited but white colonists. Thus, the Kikuyu had no reason to hope for improvement and much to fear from the twenty-nine thousand whites who dominated them.

The insurgency, known to the British as the Emergency, began in 1952, and most of the military action took place between October of that year and July 1953. It was begun by what David Anderson described as "a rag-tag band of perhaps thirty followers" led by a former soldier in Britain's King's African Rifles, Waruhiu Itote, who adopted the nom de guerre "General China." Under his leadership, the force grew to about four thousand warriors (*itungati*), whose attacks the British officers regarded as well planned and efficiently executed. In the early days of the rebellion, they virtually took over the Kikuyu part of Kenya.

The settlers were horrified to see their way of life threatened and demanded massive and merciless suppression of "the savages." The British responded both massively and without mercy. The newly arrived governor, Sir Evelyn Baring, found that his staff knew little about the Mau Mau or even what motivated the disaffected members of the Kikuyu, so he did what he was expected to do: he consulted the Kikuyu "chiefs," men who had been appointed by the British and upon whom they had relied since the formation of the colony. Naturally, the chiefs told the British what they thought the British wanted to hear and gave them such information against their rivals as would protect their own positions. Baring toured the colony to see for himself and wrote, "I've never seen such faces, they were scowling, they looked unhappy, they were intensely suspicious." The excellent American scholar on Kenya, Caroline Elkins, remarks that "there is no record of Baring ever suggesting during his introductory tour or at any time during the 'Emergency' that the Kikuyu people might have had a genuine social or economic grievance. From the time he arrived in Kenya, he accepted uncritically the notion that Mau Mau was a completely illegitimate movement." So Baring concluded that "if you don't get Kenyatta and those all around him and shut them up somehow or other we are in a terrible, hopeless position."

Baring ordered a massive arrest of suspects. But the suspects turned out to be the more moderate of the Kikuyu leaders (including

Jomo Kenyatta, who had been trying to calm the dissidents) while overlooking the less-well-known younger men. The result was that the insurgency intensified and gathered new adherents. A flow of recruits was crucial since the British killed virtually every one of the *itungati* they found. As volunteers came forward, they were trained by those members of the Mau Mau who had served in the British army. The military side of the operation—and the counterinsurgency—have received the most attention, as they have in other insurgencies. More important for the survival of the movement, however, were the politics and administration of the movement. The guerrillas "organized an intricate, passive-wing operation that would provide intelligence, weapons, food, and other supplies to the forest fighters. It was the size of this passive-wing organization that reflected the grassroots depth of the movement." To meet the Mau Mau challenge, Britain had a curious collection of paramilitary formations—in addition to a Home Guard composed of black Kenyans loyal to Britain and a regular army unit known as the King's African Rifles, composed of blacks from all over East Africa who served under British officers, there was a white settler militia and police force that together numbered about five thousand men. Given their backgrounds and the fear and hatred aroused by the Mau Mau, the settlers among these troops would attempt to exterminate all those "Kukes" who fell into their hands.

They did. Bounties were paid for dead suspects and competitions with "scoreboards" were held to see who could kill the most, with "five shillings a nob." Then, on November 23, 1952, a massacre was perpetrated in the village of Kiruara that resembled the 1919 British massacre of a similar gathering of men, women, and children at Amritsar, India. More than a hundred unarmed Kikuyu were killed while listening to a lecture in which the speaker recounted a vision he had dreamed of the end of British rule. Unlike the India massacre, the African massacre attracted little attention abroad. The bodies of the audience dead were hurriedly buried, and the official account minimized the incident.

When the civilian governor Sir Evelyn Baring was judged to have made little progress in suppressing the revolt, the British government sent to Kenya in the spring of 1953 a British general, Sir George Erskine, who took an immediate dislike both to the forces under his command and to the settler community, whose brutality and racism appalled him. He ordered the security police to stop "'beating up' the inhabitants of this country just because they are the inhabitants" and his troops "to stop at once on any conduct which he would be ashamed to see used against his own people." But Erskine was no liberal and certainly no pacifist. After studying the situation for four months, he concluded that "from evidence coming forward from screening teams and the improved intelligence services, it is now clear that Mau Mau is wider spread and deeper rooted than was thought possible even six months ago." He augmented the locally available forces by bringing in three battalions of regular British troops, an artillery battery, an armored car or light tank squadron, and a number of fighter bombers and even heavy bombers.

At first it appeared that the British government had fallen into the mental trap that is so much a feature of counterinsurgency: meet danger with force. But Erskine soon realized that force was not working. The government in London also found it prohibitively expensive. So it dispatched to Kenya a special mission headed by Hugh Fraser, a Conservative member of Parliament, to develop a new plan. Fraser's answers avoided the political question, the rule of Kenya by foreigners, but suggested a counterinsurgency strategy for defeating the Mau Mau. What Britain should do, Mr. Fraser argued, was to mount a program of "rehabilitation" that "did not rely entirely on violence and repression . . . This strategy would offer social and economic change to those Kikuyu who confessed their oaths and then cooperated with colonial authorities in the detention camps, and eventually in the Emergency villages in the Kikuyu reserves. Rehabilitation would be the inducement needed to lure the Kikuyu away from Mau Mau savagery and toward the enlightenment of Western civilization. It would offer Mau Mau adherents opportunities far more alluring than those offered by their own movement. Rehabilitation was to become the colonial government's campaign for the hearts and minds of the Kikuyu."

The phrases sounded promising but the means to implement them proved elusive. So Governor Baring began to search far afield for answers. In Malaya, he found what appeared to be a ready "school" of counterinsurgency. The British administration of that colony had been struggling against a mainly ethnically Chinese rebellion since 1948. To Baring then, and later to the Americans in Vietnam, Malaya looked like a success story. To find out how the British had apparently turned the trick in Malay, Baring dispatched a senior British Colonial Office official to Malaya. After a two-week-long briefing, Thomas Askwith returned to Kenya ready to apply the Malayan model. But the model did not fit there—as it would not fit when tried later in Vietnam. In Malaya, the British, guided by the then leading counterinsurgency guru, Robert (later Sir Robert) Thompson, were able to deport as foreigners more than fifteen thousand ethnic Chinese detainees whom they regarded as irredeemable Communists. It was hardly possible to exile even "hard-core" Mau Mau Kikuyu as non-Kenyan. Moreover, the Malays disliked the Chinese, who were even more intrusive foreigners than the British, while the insurgents in Kenya were fellow Kikuyu. So Baring and General Erskine concluded that tactics suitable to local issues were required.

Erskine decided that what he needed above all was information—intelligence—on the Mau Mau movement. Strategically, he wanted to find out why the Mau Mau fought and why the Kikuyu supported them, and tactically, he wanted to find out who the supporters were and how they supplied the Mau Mau. The means he chose was grilling, that is, questioning under torture, captured insurgents and suspected supporters. Once they had supplied what information could be extorted from them, he had those who remained alive put in detention camps, out of touch with the still active guerrillas.

By a lucky break, his forces managed to apprehend the guerrilla leader, Waruhiu Itote ("General China") in January 1954. Under intensive interrogation, he gradually revealed virtually everything the British wanted to know, including the location of his headquarters, the shape of his support organization outside the forest, and the size and structure of his guerrilla force. The guerrillas were revealed to be less than half the number the British had thought (twenty thousand). Itote put their number at seventy-five hundred. To the surprise of his interrogator, he revealed that, like the Spanish guerrillas and Tito's Partisans, the Mau Mau had created a factory to manufacture rudimentary firearms and to repair those they captured. The British obviously did not think the Kenyans capable of such efforts. Even more surprising, they found, was the pitiful armament with which they were fighting the British army to a standstill. Itote admitted to having only 14 automatic rifles, 361 bolt action rifles or shotguns, 4 grenades, 1 Bren gun, 23 revolvers, and 1,230 firearms made from ordinary water pipe, door bolts, and springs. Nothing had reached them from outside Kenya.

Despite the invaluable intelligence they got from Itote, the British were making little progress against the Mau Mau because, like all sensible guerrillas, those who remained at large fled when at a disadvantage. Even with their enormous advantages of mobility by truck and aircraft, when the British chased the insurgents, they often lost them in the forests and ravines around Mount Kenya. So the British persuaded Itote to arrange peace talks. He tried, but the talks broke down because neither side was willing to trust the other. Cleverly, the British benefited from the hiatus to identify warriors and supporters so that immediately after the truce ended, they arrested more than a thousand Kikuyu and began a massive program of detention. Effectively, they imprisoned the entire urban population of the Kikuyu and two other tribes. Whole villages were depopulated, and virtually every male African in Nairobi was separated from his wife and children, who were packed off to the reserves with no means of feeding or housing themselves. More than 30,000 people were thus swept from their homes. The British ultimately penned up about 150,000 Kenyans in concentration camps.

Having gone far to break the contact between the insurgents and their supporters, the British brought in an engineering regiment to pierce the mountain forest with roads along which their armored cars and truck-borne infantry could move rapidly. Whenever they flushed out a guerrilla group, they employed their aircraft to bomb and machine gun whomever they spotted. And the British began to employ "hunter-killer" groups, known as "pseudo gangs" of loyal blacks who pretended to be Mau Mau and so lured the real Mau Mau into ambushes. These pseudo gangs, led by white settlers, took prisoners only for the purpose of extracting information by torture before killing them. In this way about twenty thousand Kikuyu were killed.

But still, the Mau Mau did not give up, so the British brought forward two political answers to the insurgency challenge. The most obvious was to make use of the cultural and religious splits in the Kikuyu community. The older generation was apt to be influenced, they decided, by the traditional leadership, the "chiefs." Since they appointed the chiefs, the British could count on them. Moreover, as beneficiaries of British rule, the chiefs were almost as afraid of the radical change probable in a Mau Mau victory as the white settlers. So they willingly collaborated with the counterinsurgency. Second, since as a result of generations of missionary activity, large numbers of tribesmen had converted to Christianity and found the oath-taking ceremony an abomination, they feared and hated what the Mau Mau were doing even when they sympathized with the objective. On a more "practical" level, the colonial government also encouraged the feeling of loyalty to itself by the promise of land grants to those who agreed to take up arms against the Mau Mau. In this way, the government was able to get the chiefs to organize the Christians into private home guard units. Within a year, these forces were at least three times as large as the Mau Mau.

But still the insurgency did not stop. It began to be clear to the British authorities that two problems shaped the Kenyan experience, land and the oath. To these issues, the British developed three answers. The first was an attempt to meet the desperate hunger for land among the Kikuyu, the second was to find a way to release the Mau Mau from their oaths, and the third was to bring forward an acceptable leader in the place of the formerly militant "General China." The two key shapers of these related policies were Thomas Askwith and Louis Leakey. They were as crucial to the suppression of the Kenyan insurgency as Sir Robert Thompson, Edward Lansdale, and John Vance were later to the counterinsurgency policies in Vietnam. But, whereas Thompson, Lansdale, and Vance failed in Vietnam, Askwith and Leakey largely succeeded in Kenya. So how did they do it? First consider land.

The bitterest issue among the Kikuyu was the British confiscation of the tribal lands. That, together with the rise of population after the First World War, turned large numbers of Kikuyu into landless laborers. Then, as I have shown, in the 1930s additional regulations were implemented that swelled the total still further. Those people who had lost their lands in the "White Highlands" had no hope of finding lands elsewhere, because everywhere else in Kenya overcrowding was prevalent. Consequently, people flocked into the towns and particularly into Nairobi. In the new slums, the then idle landless farmers had no trade or skill to sustain themselves. As Askwith realized, if Kenya was to achieve peace, this problem had to be addressed, but Governor Baring was adamant that the rebels not be "rewarded" for their violence by being given lands. He favored increasing the productivity of existing lands, a policy that not only was probably unworkable given the restraints of technology, money, and social customs, but had the effect of favoring the landed rich without addressing the problem of the landless poor. As a result of his opposition to any form of large-scale land distribution, more than a million Kikuyu were crammed into government-run "villages." These villages were Kenya's version of the contemporary fortified village program of the British in Malaya and the later "strategic hamlet" program of America in Vietnam. The inhabitants regarded them as prison camps and hated them. Even if they had been acceptable, however, they made no economic sense to the inhabitants because the lands surrounding them were of poor quality. The only source of good agricultural land was the white settler community. Ultimately, a repurchase of settler land was forced on the reluctant governor.

The second problem to be solved in order to bring the insurgency to a halt or at least to diminish its impact was the oath. Advised by the well-known paleoanthropologist Louis Leakey on the nature and importance of oath taking, the colonial government commissioned him to work out a "counter-oath" ceremony to "cleanse" those who had taken it and so wean them away from the movement. Having grown up in Kenya, Leakey prided himself on being, as he titled one of his books, a *White African*. A devout Christian, he believed that the only effective counterinsurgency weapon was Christianity. And with his knowledge of Kikuyu culture, he was aware that there was an "unoathing" ceremony that would be far more effective than simple repression. It was certainly the most penetrating application of counterinsurgency ever devised. Under Leakey's inspiration, the British began encouraging defections from the Mau Mau movement with the promise of "rehabilitation." But the statistics indicate that it often did not work. For those who did not take this option, long-term imprisonment or death by hanging was the alternative. After what were usually perfunctory trials of those suspected of working with the Mau Mau, the British hanged 1,090 Kikuyu. That number, as David Anderson points out, is double the number of convicted terrorists executed by the French in Algeria and many more than the combined total for the insurgencies in Aden, Cyprus, Malaya, and the Palestine Mandate. "In no other place, and at no other time, was state execution used on such a scale as this . . . British justice in 1950s Kenya," he wrote, "was a blunt, brutal and unsophisticated instrument of oppression."

Yet it was the settlers who actually lost the war. Why? The reason is essentially that *military and security forces cannot recreate the pre-insurgency situation.* Hanging the leaders, killing the soldiers, and penning up their supporters in concentration camps created a situation that could not be maintained. It was then, and always is, politics, not force, that counts. The settlers had begun to find their lives unprofitable and unattractive, while the British government would not condone their actions to try to regain what they had lost. Although cheap by later standards in Vietnam and Iraq, the Mau Mau war was a serious drain on the British treasury. Worse, like the Algerian war for the French, it was a drain on Britain's standing in the world. So beginning in 1959 in the Conservative government of Harold Macmillan, the secretary of state for the colonies, Iain MacLeod, decided to "wipe the slate clean."

First MacLeod set about dismantling the "legal framework of the Emergency." Then he closed the concentration camps and sent the inmates home. And finally he pulled the rug out from under the white supremacists: Kenya would have majority rule with open franchise, and the white settlers would be forced to sell their lands. It was at that point in the summer of 1961 that the British brought out of prison the moderate leader, Jomo Kenyatta, to begin the task of national reconciliation that would take Kenya to national independence in 1963. In that task it was Kenyatta rather than the white colonists, the security forces, or the British troops who subdued the Mau Mau. With British help, he was able to give his people what the Mau Mau had fought for: independence. With that, at least a partial victory, General China settled down to write his memoirs, Kenyatta became president, and the Kikuyu stopped supporting the insurgency. It simply faded away.

CHAPTER 8

THE ALGERIAN WAR OF NATIONAL INDEPENDENCE

LIKE THE IRISH, THE ALGERIANS FOUGHT AN INTERMITTENT GUER-rilla war for generations against a foreign power; the foreign power in Ireland was Britain while in Algeria it was France. Both powers had managed to create sizable colonies of expatriates. In Ireland, as we have seen, they were English and Scots (who became known in America as the Scotch-Irish); in Algeria the *colons* (also known as the *pieds noirs*) were from all over the northern side of the Mediter-ranean—Spaniards, Italians, Alsatians, Corsicans—with only about one in five French. In both Ireland and Algeria, the foreigners dif-fered from the natives in religion. The Scots and English were almost entirely Protestant, while the Irish were mainly Catholic; in Algeria, the *colons* were either Christian or Jewish, whereas the natives were Muslim. In short, Algeria, like Ireland, Palestine, South Africa, and Russian Central Asia, was not so much an *imperial* as a *colonial* ven-ture.

But the war did not begin with colonialism. When the French invaded in 1830, their motivation was imperialism—partly strategy and partly greed. In addition to offsetting Britain and Spain on the North African shore, they wanted to seize the huge horde of gold held in Algiers by the then ruler. At first it seemed easy: they put their relatively modern armada of more than six hundred ships into position off the city of Algiers and, before an appreciative French audience that had sailed across the Mediterranean in small yachts, bombarded the city and sent a thirty-thousand-man army ashore. In four weeks, they defeated the antiquated Algerian army. But then, after the initial "shock and awe," they stumbled into a guerrilla war.

Like many Middle Eastern and African countries, "Algeria" was a flexible concept. The Arabic name recognizes this by using the plural: *al-jazair* means "the islands," which to the Arabs meant inhabited places in sea or sand. The country was an extension of the principal city, Algiers; how much territory it controlled varied according to the power of the ruler. As we know it today, it is vast. The second largest country in Africa, it is about the size of the American Great Plains states. But much of this area is desert—even today with sophisticated irrigation, only about 3 percent is arable. In former times, the vast interior beyond the Atlas mountains was populated in the only way it could be, by villagers on scattered oases and nomadic tribesmen who moved to feed their animals according to the chance fall of rain. City men rarely penetrated deep into the interior. So the tribal areas and oases were virtually separate countries. Historically, Algeria was effectively just the coast and the narrow inland plain where rainfall made agriculture possible.

Within these restraints, Algeria made itself an important country in the seventeenth, eighteenth, and early nineteenth centuries. It did so primarily by turning to the sea. In the long wars with Spain, France, and England, its navy did exactly what the fleets of those countries did—engaged in a mixture of trade and privateering. So common was this practice that it was regularized in treaties with a set system for recycling cargoes and ransoming prisoners. Today we would be tempted to think of both the Europeans and North Africans as rogue states and their sailors as terrorists. But that was not how contemporaries regarded one another. They were engaged in what amounted to a cold war with occasional bouts of fighting interspersed with long periods of commerce.

The two sides of the Mediterranean were only partly divided by religion, with the Europeans Catholic Christians and the North Africans Sunni Muslims. Apparently no Muslims then lived in Europe, but many Europeans worked and fought for the North Africans. In the eighteenth century, some 16,000 of the 120,000 inhabitants of Algiers were Christian *renegados,* as defectors from Europe and England were called, and they supplied a significant portion of the corsair crews. That fleet was the industry of Algiers. On land, shipbuilding and trade in shipborne commerce were the main occupations of the diverse population.

In addition to the Europeans, there were descendants of the people the Romans knew as Numidians and whom we call Berbers; most of the Berbers had been Arabized and virtually all converted to Islam during the eighth century A.D. Large numbers of Spanish Muslims (Moriscos) and Sephardic Jews crossed into North Africa after their expulsion from Spain in 1492. Unlike the stern and intolerant Christians of Castile and Aragon, the Muslims of the North African city-states welcomed the Jews. By the nineteenth century the thriving Jewish community of Algiers was nearly as numerous as the Christian *renegados.* They were not the only foreigners. Since the absorption of Algeria into the Ottoman Empire from 1518, a steady stream of Turks had arrived as soldiers and administrators. So, unlike the southern European states which generally persecuted or expelled not only Muslims and Jews but also Protestant Christians, Algiers was a cosmopolitan society in which each group was allowed to practice its own religion and follow its own customs. There, even prisoners who were Christians were allowed a church in their jail.

The administration of Algiers was much like contemporary Venice and other European city-states, except that the Ottoman Turks were the elite group and supplied the rulers. The merchant community, however, wielded a powerful influence that balanced the power of the soldiers. Viewing it from a distance, Europeans denigrated Algiers as "a turbulent state, forever on the brink of anarchy and rebellion [but] the fact is that during the Regency's [the local government's] three centuries of existence there was not a single large-scale revolt against the central authority." The hand of the Turks generally rested fairly lightly on the population. It had to be restrained for the city to generate the prosperity European merchants and diplomats found there.

Napoleon was one who found its prosperity important. During his voyage to the East in 1798, he arranged to buy grain, which was then one of the major exports of Algiers, for the thirty-eight thousand men he was taking to Egypt. The Algerians were happy to supply it, but they were not happy when Napoleon defaulted on payment. His failure to pay began years of legal and diplomatic negotiations, at the end of which, in 1827, the exasperated ruler, furious that he had been swindled, lost his temper and flicked the resident French consul with his fly whisk. When the incident was reported to the French government, it demanded a particularly humiliating form of apology, which the Algerian ruler refused, saying to another European consul, "I'm surprised that the French haven't asked for my wife as well." Then, having threatened dire punishment, but doing little, the French resorted to what we call sanctions: they blockaded the port. Two years passed during which an unpopular French government tried to whip up domestic public support by proclaiming a foreign policy aimed at "revenge." Retaliation was the excuse for the French attack in 1830.

The French invasion at first went smoothly. It began almost as a gala affair. Wilfred Blunt painted a verbal picture of the event. "In the bright sunlight of these May mornings the whole venture wore the aspect of a vast, care-free pageant. Official painters lugged their easels and canvasses on board [the French warships]; representatives of foreign powers, who had joined the expedition to see the fun, strutted about in gay, outlandish uniforms . . ." But as so often occurs in picnics, the fun was dampened by bad weather. The gaudy ships were forced to retreat to the island of Majorca, where half the pride of the French navy became immobilized by engine failure. It was a portent of troubles to come. But no one was in a mood to read the portents. The expedition returned to the African coast where, with only sporadic opposition, the troops landed. In their first encounter, they overwhelmed the Algerians, who lost all their artillery and some four thousand men while the French lost only fifty-seven killed and five hundred wounded. It was a battle typical of the nineteenth-century clash of Europe and Africa. But it did not end the war.

As Blunt wrote of the French in Algeria, many would later write of the Americans in Iraq: "A sad disillusionment lay ahead . . . Rarely had a conquering nation known so little of the peoples, the customs, and even the elementary geography of the land it imagined it had virtually subjugated." When the ruler surrendered, the French proclaimed that their mission was accomplished. In a grand ceremony, they allowed the ruler to go into exile with a small part of his treasury and all of his harem. But other Algerians refused to recognize their defeat. Angry that the Algerians continued fighting, the French looted Algiers city and began attacks on outlying villages. Each attack tended to galvanize opposition rather than pacify people. The resistance began to take shape.

Resistance to the French invasion was led by one of the most remarkable men of the nineteenth century, the then twenty-two-year-old Abdul-Qadir (often spelled Abd el Kader), a religious figure who based his call to arms on religion rather than nationalism. Indeed, at that point in his life, neither he nor his followers would probably have understood "nationalism," for which Arabic did not then have a word. Abdul-Qadir based himself in the town of Mascara. The town was well named: in Algerian Arabic *mascara* meant "mother of soldiers." From it Abdul-Qadir rallied townsmen and tribesmen from all over Algeria to fight the French invaders every step they took for fifteen years.

Abdul-Qadir, whom the foremost English student of colonial wars called "a partisan leader of the foremost rank," realized that he could not hope to defeat the French in a mass battle. As Colonel (later Major General Sir) Charles Collwell wrote, "His personal ascendancy over the nomads of the south and over the wild Berber hill-men, enabled him to play the strategist from wherever his wandering abode might be, over a vast extent of country. For years his bold and sudden strokes bewildered the French leaders. He had full information of their slow, deliberate movements. He knew when a garrison was drained of troops and he straightway fell upon it. He cut communications and he swept away convoys. But by the time the French had assembled their forces for a counter-stroke, and had dragged their guns and impedimenta to the spot, the hostile body which had done the mischief had vanished into the desert, or had flown to some hill refuge whither the cumbrous column could not follow it."

Against Abdul-Qadir, the French brought a general who knew guerrilla warfare firsthand. Thomas Bugeaud had studied it the hard way. As a young officer in 1808, he was in the first French column to enter Spain and experienced the Spanish *guerrilla* both in the Second of May uprising in Madrid and in the second siege of Saragossa. From Spain, as Collwell continued, "He perceived that he had to deal not with a hostile army but with a hostile population, that this population consisted largely of clans and tribes of fixed abode, and that to bring them to reason he must reach [them] through their crops, their flocks, and their property." So he set up "flying columns" whose task was "to chastise the rebels in their homes . . ." He was the first of France's several ruthless military commanders who would beat the Algerians down into defeat.

As the French took town after town and cut Abdul-Qadir's forces off from his people, Abdul-Qadir created a "nation-in-motion" much as Tito was to do a century later in Yugoslavia. Later he told a French general, "My *smala* [camp, which at times numbered sixty thousand or seventy thousand people] included every kind of craftsman necessary for our organization—armourers, tailors, saddlers and the like. Complete order prevailed. *Kadis* administered justice, exactly as they did in the towns. Markets were held. There was no stealing, no immorality. Hospitality was offered, as in the past. When we made a halt, the education of our children went on; the times of prayer were observed, and announced by the *muezzins*. Each family carried provisions according to its transport facilities, the rich for two or three months, the poor for at least a fortnight."

Abdul-Qadir fought on, but the French tactics gradually drained the popular support on which his warriors depended, and the frightful suffering caused by Bugeaud's attacks on their villages disheartened even his warriors. Then he was misled by a chance victory at the battle of Macta on July 26, 1835, into making the classical mistake of resistance leaders: he tried to transform his guerrilla forces into an army; this force, losing some of its mobility and offering a better target, could not match French numbers—which rose to more than one hundred thousand men, a third of the whole French army. In a series of desperate evasions and pitched battles, Abdul-Qadir fought and ran in a way that almost perfectly paralleled the furious campaign Tito waged against the German army. But he lacked Tito's system of control, the ideology that unified his followers, and the ruthlessness that punished defection. One after another, Abdul-Qadir's tribal allies defected. In 1846, with his army worn and reduced to a remnant, and unwilling to bring further suffering on his people, he surrendered.

Meanwhile, Europeans poured into the country. Within a decade after the French invasion, Europeans amounted to about 1 percent of the population and were hurriedly taking over the agricultural lands. As in the Philippine and Vietnamese insurgencies, ownership of land was bitterly contested, but it was recognized as key to winning the war. The French strategist Thomas Bugeaud seized upon the critical issue: "Wherever there is fresh water and fertile land, we must plant colonists." Colonialism was seen as the ultimate form of counterinsurgency: what they could not effect with the rifle, the French thought they might bring about with the plow. Large areas were confiscated officially in reprisal for resistance, and the inhabitants were either killed or driven away. As in contemporary America, where rights to Indian lands were being, in the current phrase, "extinguished," there were occasional efforts to slow down or at least regularize seizure of lands, but these efforts were easily circumvented. Ironically, it was Napoleon III, not a ruler noted for his concern for France's overseas subject population, who arranged the passage of a law in 1863 that recognized "the ownership of Arab tribes of territories of which they have permanent and traditional benefit." Despite this, in 1872, in one of many transactions, the French forced the Berbers of the eastern province to give up 453,000 hectares (1,750 square miles). Area by area the Algerian population was driven out of virtually all the usable agricultural lands.

The colonists not only took away the land of the natives but drove them down culturally. When the French invaded, a large portion of the Algerian population was literate in Arabic, with a satisfactory standard of living and a coherent cultural life shaped by Islam. But already by 1847, Alexis de Tocqueville told his colleagues in the French National Assembly that "we have rendered Muslim society much more miserable and much more barbaric than it was before it became acquainted with us." It got much worse. After a century of French rule, about three in each four Algerians were illiterate even in Arabic, few had stable jobs, and almost none had well-paying employment. When I went to Algeria in 1962 in the last days of the Algerian-French war (which I had been trying to help stop on behalf of the American government), I found that Algerians were so totally excluded from the *colon* economy that even "mom and pop" laundries and bakeries were European monopolies. Although Algiers had France's largest medical school/hospital complex, fewer than half a dozen of its doctors were Algerian. There was no foothold for an Algerian in *colon* Algeria. Among the excluded natives, bad health was chronic and many went to bed each night hungry.

In a convoluted political structure, France organized Algeria *both* as a colony, in which the vast majority of the population was not accorded French citizenship, *and* from 1848 as an integral part of France, divided into three *départements* (administrative provinces) similar to those in Metropolitan France. While legally a part of France, the administration was kept separate from the other French *départements*, and the army played a role there that would not have been tolerated in Metropolitan France.

Until the end of the First World War, the Algerians were just victims. They played no part in administering themselves, and their voices were effectively silenced. Then, despite the bitter opposition of the *colons*, the National Assembly in Paris recognized their contribution to the war effort—in which 173,000 Algerians had enlisted in the French army and nearly one in each six was killed—by moving slowly and haltingly toward a policy of turning them first culturally and then politically into Frenchmen. As this policy progressed, a socialist government in Paris in the mid 1930s offered citizenship to a select group—one Algerian in each 250. The *colons* were outraged and organized to overturn the legislation. Its principal sponsor, Maurice Viollette, presciently warned his parliamentary colleagues that if the Algerians were not allowed to become members of the French nation, "beware lest they do not soon create one for themselves." He was far in advance of his time and was virtually laughed out of the Assembly. Meanwhile, not yet attracted to nationalism, Algerians sought refuge in Islam.

While Islam offered a unifying code of conduct, it was not able to overcome the ethnic divisions of Arab and Berber, the social divisions of tribes in each ethnic group, or the cultural divide between villagers and the urban population. These divisions existed before the arrival of the French and contributed to the failure of Abdul-Qadir's attempt to prevent the conquest. In rural areas, peasants and tribesmen thought of their little worlds, their villages, rather than Algeria as their nation. Between rural villager and townsmen, there was little affinity. What gave them a semblance of unity was that all were despised by the Europeans. Both the peasants, who were treated as a species of farm animal, and the "street Arabs," who were thought of as petty thieves, were circumscribed in residence and movement and were not allowed in higher education or most professions. For many years, all suffered in silence, but gradually they sought means of change.

The means of change grew slowly and in some confusion. Three main Algerian resistance groups evolved but never found common ground. The first, based on religion, had a long history abroad and became an important factor in Algeria in 1931. It was the Jamiyah al-Ulama al-Jazariyah (the Union of Muslim Learned Men of Algeria), which was originally an association of Muslim clerics who were influenced by similar revivalist movements in Egypt and Morocco.

Somewhat like the Sunni Hamas in modern Palestine and Shia Hizbullah in modern Lebanon, the Jamiyah drew strength from the social services it provided for the native population; however, it never managed to develop an effective political program.

The second major group, which after several aborted trials formed the Union Démocratique du Manifeste Algérien, was organized by the small urban elite of relatively prosperous and Europeanized Algerians. Its members shared the French view of Algeria: it was a backward society. The way forward, they argued, was to put aside Algerian culture and evolve—hence they were known as the *évolués*—from "backwardness" toward French culture and French citizenship. Like most of the leading Americans on the eve of their Revolution, the Algerian Unionists sought not independence but equality. The man who was their recognized leader, Farhat Abbas, famously denied that there was such an entity as the Algerian nation. Although he had enlisted in the French army, was married to a French woman, was an *évolué* as a educated man (as a pharmacist), and was subsequently elected to the Assembly, he was effectively marginalized as the "token Algerian."

The third group was the most radical. It was organized and led by an Algerian of humble origin who had fought in the French army during the First World War and had married a French woman. Although Messali Hadj had practically no formal education, Algerians found him a spellbinding public speaker. His first political organizing was among the Algerians working in France who were underpaid, discriminated against, and angry. He offered them a means of expressing their anger in a sort of political club (for which the French authorities put him in prison); perhaps more important and certainly more popular, in his personal deportment and even in his dress he offered these despised "reverse-*colons*"—Algerians living in France—enhanced self-esteem. He became the visible symbol of Algeria. In 1937 he enlarged his political club into a real political party, the Parti Progressiste Algérien, and gave it a full-blown political and social program including both independence from France and redistribution of the lands then held by the *colons*. During the Second

World War, Messali was condemned to sixteen years of hard labor and the party was banned.

Meanwhile, during the dreary years of the Second World War, when Algeria was under Vichy control, the Algerians, like virtually all the inhabitants of the African and Asian nations, listened avidly when Roosevelt and Churchill made their grand pronouncements about freedom and independence. Not to be overshadowed, Charles de Gaulle faintly echoed their words in January 1944 when, on behalf of the Free French, he cautiously and vaguely promised "to lead each of the colonial peoples to a development that will permit them to administer themselves, and, later, to govern themselves."

The Algerians, to their cost, believed Roosevelt and Churchill and heard what they wanted to hear in de Gaulle's statement. So, on the morning of May 8, 1945—VE day—inhabitants of the little Algerian town of Sétif gathered to celebrate. They thought, naïvely to be sure, that the millennium had arrived. Joy got out of hand and celebration turned to riot as the participants waved the flag of the early resistance leader, Abdul-Qadir. Particularly among the landless agricultural workers who still lived "in a state close to peonage," momentary joy quickly was overwhelmed by pent-up anger. Ugly but sporadic attacks on Europeans drew down upon them the fury of the *colons.* Between ten thousand and forty-five thousand Algerians were massacred by private Frenchmen, the French police, and the French army. Even distant villages, some forty of them that were inhabited by Arabs, were bombed by aircraft that had been supplied to the French to fight the Germans. That tragedy may be taken as the seedbed of modern Algerian nationalism.

The French Communist Party, to which the Algerian radicals had looked for at least psychological support, dumped them just as its Moscow bosses were in the process of doing to the Greek resistance. Playing European politics, where it wanted to influence the decisions on the move toward the creation of NATO, it was showing its French patriotism by coming out in opposition to Algerian independence. Its newspaper, *Humanité,* downplayed the Sétif massacre, while its Algerian offshoot, *Liberté,* urged that the Algerians who had instigated the revolt be put in front of a firing squad. The Communists showed that they were above all else Europeans. Allegedly, some Communists took part in the reprisals, and even as late as 1954 the Algerian Party, in which most members were *pieds noirs,* supported the sending of French soldiers to fight the Algerians.

The Sétif massacre and the Communist reaction to it caused Messali Hadj to break with the Communist-led "Front" and to reconstitute his suppressed party as the Mouvement pour le Triomphe des Libertés Démocratiques. It won the municipal elections of 1947 but was overwhelmed by fraud and intimidation in the crucial 1948 elections for the Assembly. Following that defeat, Messali Hadj was again arrested and quickly deported. The net result of his fall was that a new generation of Algerians, many of whom had served in the French army during the Second World War, concluded that they could gain nothing with ballots and began to think in terms of bullets.

Among the young veterans was Ahmad Ben Bella, who had won two of France's highest decorations for valor—one of which was awarded personally by Charles de Gaulle. As a soldier, Ben Bella had fought in the Italian campaign; it was there that he first encountered and was deeply impressed by a resistance movement. Upon his demobilization and return to Algeria, he had entered politics as a follower of Messali Hadj. He was successful on a local level but quickly sensed that he could not cross the barrier constructed to keep Algerians out of major positions. Breaking with Messali Hadj, he formed a more militant offshoot known as l'Organisation Spéciale (OS) to fight for independence. In need of funds for political action, he helped to organize the robbery of a post office. He was caught and sentenced to eight years' imprisonment. The OS, which by then had grown to nearly five thousand, was broken up; most of the key men, several of whom would later become his colleagues in the resistance, fled abroad. Ben Bella managed to escape and went underground as he had seen the anti-Fascist Italians do.

By then the Algerian Communist Party had realized that its stance favoring French rule had cost it its influence; so from the profoundly anti-nationalist position it had advocated after the 1945 Sétif massacres, its leaders called for a nationalist "Front" with the other parties. This was all the proof the increasingly worried *colons* needed to charge that *all* the nationalists were Communists. Far from it, they were almost as bitterly split among themselves as they were hostile to French rule. As they proved unable to coalesce in the nationalist cause, a group of the younger radicals decided to form yet another new organization to fight for independence.

Meanwhile, far from Algeria, two things happened that would shape that fight. First, the French army was defeated by the Viet Minh at Dien Bien Phu. That great battle was the first time that a French army had been defeated by a colonial people. The French defeat had the same effect in Algeria that the defeat of General Braddock's British army during the French and Indian war in 1763 had on the colonial Americans: it convinced Algerians that they too might be able to defeat the French.

The second result of the French defeat in Vietnam was that a group of French army officers, having been humiliated by the Germans in the Second World War and by the Viet Minh in Indochina, came to share an almost mystical vision of a program aimed at recapturing French grandeur. What these men believed had made the Viet Minh able to defeat them would give rise to an entirely new form of counterinsurgency. Politically sensitive, utterly ruthless, and sharply targeted, French military counterinsurgency would be first played out in Algeria by the French elite paratroop regiments under General Jacques Massu and later attempted by the Americans in Vietnam under Robert Komer, William Colby, and John Paul Vann.

On November 1, 1954, after a great deal of preliminary jockeying for position, the Algerians who felt most keenly their national destiny formed the Front de Libération Nationale, the FLN. The name testified how completely France had wiped away the cultural heritage of Algeria: even the most determined militants spoke French by preference. Years later, one of them, Houari Boumedienne, told President Nasser of Egypt (who told me) that he was the only one in the Algerian cabinet who really knew Arabic. Despite, but also in part because of, the denationalization of Algeria and Algerians, the war of independence had begun.

This, the second, or twentieth-century, phase of the war for Algeria was started by dispersed, poor, and ill-armed groups. Like the earlier American colonists at Bunker Hill and the Mau Mau guerrillas in Kenya, few Algerian would-be insurgents even had weapons. Among the widely scattered groups they had fewer than four hundred firearms. Of these about half were shotguns. (The quest for weapons was ever to be one of the determinants in their military operations.) Militarily the insurgents were insignificant, as almost all guerrilla movements are in their infancy, but psychologically they had profited from the defining moments of Sétif and Dien Bien Phu. Then, as each incident, each stirring demonstration, each infuriating humiliation, each vicious suppression piled upon others, they slowly gathered strength. But their strength was never military. They were never able to stand up to the French army, which was massively armed by the United States and which outnumbered them about forty to one. Nor were they able, like Tito's Yugoslav Partisans or the Greek EAM, to hold territory. But they began to create what I have called the climate of insurgency, the step necessary to build popular support. In an attack on the *colons* in the area around the town of Philippeville in eastern Algeria, they provoked a furious bloodbath that drew the lines indelibly between them and the French. Provoked, the French army carried out a second massacre. At Philippeville French soldiers had orders to kill every Arab they met. They killed perhaps twelve thousand. After Philippeville, there was no middle ground. Algerians were on one side and French on the other. So insurgent forces expanded from the few dozen in 1954 to perhaps twenty thousand by 1956. And, as they were increasingly intent to have happen, their struggle was also beginning to get international recognition.

In short, what up to 1954 was essentially a terrorist campaign was being hammered by the French into a politico-military guerrilla war. And perhaps ultimately even more important, Algerian actions were forcing the French into a counterinsurgency that was ugly, repressive, and unsuccessful. A French member of the governor-general's own cabinet resigned after charging that "arbitrary arrests are increasing; the so-called 'lodgement camps' [actually concentration camps] are filling up with more and more innocents . . . the army destroys, bombards, summarily executes." Such tactics ultimately so disgusted the French public and so threatened their civic order that France would lose the war more in France than in Algeria.

Meanwhile, unlike the resistance movements in Yugoslavia and Greece but similar to those in Spain and the Philippines, the Algerians were divided into self-contained groups that fought quite different campaigns with different methods.

The first of these campaigns took place in the countryside; it was the "internal" or as it came to be known the *wilayah* (district) war. It was a true guerrilla war. Particularly in the scrub and forest of the Kabylie, small groups of local men, just like the early nineteenth-century Spanish guerrillas, ambushed French troops whose heavy equipment confined them to the roads, seized what arms and supplies they could, and fled. The leaders of these attacks, two Kabyles, Ramdane Abane and Belkacem Krim, virtually picked up the struggle against France from where Abdul-Qadir had left it a century before. Building on a sequence of small military successes, they began to reach out to Algerians with a radio program called Voix de l'Algérie (the Voice of Algeria) and even a newspaper called *El Moudjahid* (*The Warrior*), but they never put aside terror as a tactic to compel the villagers to support them and to disrupt the French-appointed local administration. Like their counterparts in Yugoslavia, Vietnam, and Iraq, they realized that a more important target than the foreigner, who after all would ultimately leave, was the native who might take his place. In the first three years of the war, the *militants* killed more than seven thousand fellow Algerians, or almost seven times as many Algerians as Frenchmen. Some of these assassinations were used as an indoctrination ritual that, like the Mau Mau "oathing," was meant to convert an untested recruit into an accepted warrior.

The men on the ground in Algeria naturally felt that they were carrying the dangerous and difficult part of the resistance and tended to criticize those who had fled abroad with not doing enough to support them. Relations reached such a vitriolic point in the spring of 1956 that the "interior" leader Ramdane Abane wrote to Ahmad Ben Bella, who was then living in Cairo under the patronage of President Nasser, saying, "If you cannot do anything for us outside, come back and *die* with us. Come and fight. Otherwise consider yourselves as traitors!"

To avoid the breakup of the movement, the leaders decided to hold a conference. Abane could not leave Algeria, so Ben Bella agreed that the meeting be held there. The least dangerous place was in the Kabylie, which the French believed they had "pacified," but where the terrain offered some cover and the local guerrillas afforded some protection. Although he was invited, Ben Bella, waiting in Tripoli for some means of transport, missed the meeting, so it was essentially a gathering only of the "interior" leaders.

Led by Abane, the conferees worked from August 20, 1956, for about two weeks of intensive sessions to "regularize" the resistance. They decided that there could be no compromise and no negotiations before Algerian independence was assured. To carry forward the fight, Algeria was divided into six regions, but some still had too few militants even to count as guerrillas. But, as the forces grew larger, they were to be organized as though in a regular army. It is striking that even at that early stage of their insurgency, the Algerians, like the leaders of the guerrilla forces I have described in America, Spain, the Philippines, Yugoslavia, and Greece, sought to replace the (successful) informality of the guerrilla for the (usually ineffective) organization of a regular army. The tactical group, they decided, was to be a detachment (a *katibah*) of approximately a hundred men. They wanted to amalgamate or at least to coordinate these into larger formations, but that was seldom possible in Algeria. Even the *katibahs* proved to be too large for effective and relatively safe action; most operations took place with much smaller numbers. Never did the internal guerrilla force reach the size and organizational structure of Tito's Partisans in Yugoslavia or the ELAS in Greece.

The most important result of the Kabylie meeting was political. Just as Tito's followers convened a national assembly at the little

Bosnian town of Jajce which they temporarily held in November 1942, so the Algerians sought to formulate a national government in a clearing in the Kabylie. They created, at least on paper, a nation-wide seventeen-member congress they called La Conseil National de la Révolution Algériene, which was to be effectively run by a five-member Comité de Coordination et d'Exécution. The leaders feared, above all, disunity, but they lacked the party structure that gave Tito control; so their struggle for unity was mainly negative: they excluded the Communists. They even categorized their erst-while mentor, Messali Hadj, much as Tito regarded Mihailovic and the Greek EAM regarded Zervas.

Then, having set out an organizational structure, Ramdane Abane and his colleagues laid out a strategic plan: its centerpiece was what came to be called the battle of Algiers. After Sétif and Philippe-ville, the battle of Algiers was to be the third great defining event of the Algerian war for independence.

Up to that point, Algiers had stood apart from the guerrilla war that was fought in the countryside. However, as Ramdane Abane realized, the urban battle had to be fought if Algeria was ever to become independent. That battle could only be in the great city of Algiers with its huge native quarter, the Casbah. Only there could the FLN make a statement proclaiming its movement as the only le-gitimate representative of the Algerian people that would be heard by the whole world. To do this, Abane had to violate the fundamental law of insurgency: the requirement for mobility. It was a tremendous gamble, but he felt that it had to be taken regardless of the almost certain great loss of life.

The FLN in Algiers, unlike the guerrilla forces in the country-side, was organized in terrorist cells. Each cell was composed of not more than half a dozen men who were either closely related or were neighbors. Theoretically at least, only one of these men would be in contact with a member of another group in a sort of pyramid. In this manner, it was supposed that even if the organization lost a cell, the overall structure would remain secure. The "pyramid" was the classic model of a terrorist organization. But it was not so much for terrorism as for communications that Abane sought to use it. The one hundred thousand inhabitants of Algiers had joined in proclaiming to the world that Algeria was determined to be independent. Their proclamation would be a general strike of the Casbah. It was to begin on January 28, 1957.

The French governor-general, Robert Lacoste, understood exactly what the FLN aimed to accomplish and decided that if Algeria was to be "saved," the strike had to be broken. So, on his own authority, he ordered the commander of the French paratroopers—the only part of the largely conscript French army on which he could rely—to break the strike swiftly and regardless of the cost. Lacoste's was a command that General Jacques Massau welcomed. Still stinging from the French defeats in the Second World War and Indochina and the abortive Anglo-French-Israeli 1956 invasion of Egypt, he threw his paratroopers into the urban battle. The "paras" ripped the shutters off shops, drove young men found on the streets into work stations, and even rounded up students and forced them back into schools. Then they began the real crackdown. Arresting men more or less indiscriminately, they began to apply torture, which the French had used for many years throughout North Africa. The aim was to extract information on the FLN underground. Cell by cell they broke into the "pyramid" and managed to trap and kill one after another the operational chiefs. By the time they had finished, the urban underground was wiped out. Militarily, the battle of Algiers was a French victory. Politically, it was a disaster. The hideous record of torture gave rise to a sense of revulsion among the French public. And, feeling sold out by the French public, the army turned on France itself. It was the threat to France that in 1962 would drive General de Gaulle to give up.

Only de Gaulle among the French leaders was strong enough to take that painful decision. And it almost cost de Gaulle his life. He was marked for assassination by *colons,* who in 1960 had formed their own terrorist group, the Secret Army Organization (l'Organisation de l'Armée Secrète, or OAS), which in 1961 and 1962 was literally blowing up Algeria—the French coined a new word, *plastiquer,* to mean setting off plastic bombs. Even more troublesome to the government in Paris, the senior officers and many of the professional soldiers of the French army serving in Algeria revolted and threatened to invade France. After ringing Paris with anti-aircraft cannon and the National Assembly with tanks, de Gaulle faced down the revolt and rallied the army, which, following a temporary cease-fire with the Algerians, fought a second battle of Algiers—sending twenty thousand French soldiers armed with tanks, artillery, and aircraft against the *colons* who were aiding the OAS. That second battle of Algiers was a stunning example of the final cost of insurgency: the threat to the civic institutions of the dominant power.

Meanwhile, driven by the ruthless campaign against the internal army and the insurgents in Algiers, Ben Bella and his remaining colleagues had formed an "external" army in Tunisia. Eventually numbering 150,000 men, the army and its dependents were sealed off from Algeria in 1957 by a barrier known as the Morice Line (rivaling the Berlin Wall with mine fields, electrified wire emplacements, and sophisticated radar warning systems) along the whole length of the Tunisian-Algerian frontier. The external army suffered no casualties and fought no battles but was armed with the full range of modern military equipment. So, when the French finally agreed to Algerian independence in the Evian Accords on March 17, 1962, the "external" FLN was in position to dominate the relatively tiny and scattered groups of "internal" guerrillas. They could do so not only militarily but also politically because the Algerian public felt that the aim they had struggled to achieve in nearly eight years of suffering had been accomplished—the French were leaving—and they were tired. In Algeria as in Ireland under de Valera, the guerrillas had become superfluous. And, as in Ireland, natives could do what foreigners were unable to do, suppress the guerrillas. The incoming president Ben Bella quickly did so.

The cost of the war was horrific and in some ways is still being paid. During the eight years of fighting, at least half a million Algerians—about one in each sixteen natives—and about twenty-five thousand French troops were killed in combat. Tens of thousands of Algerians were locked away in concentration camps. One camp I saw when I toured Algeria was still filled with hundreds of what the French called "street Arabs," orphan children. About 1.2 million Europeans fled or were driven out of the country along with about two hundred thousand Algerian "Loyalists." Far from the battle, France itself was severely damaged. The much vaunted French *civilisation* was corrupted by the horrors of the conflict. In the memorable phrase of the time, torture had become "the cancer of democracy." Hate, fear, and shame overcame everything. Even so civilized a man as the writer Albert Camus became an apologist for French brutality. Among the most vicious of the paratroopers were men who had suffered in Nazi concentration camps or had fought in the French resistance, yet saw no moral inconsistency between their struggle for freedom in France and torturing or murdering Algerians. Civilization itself was a victim of the war. The costs have yet to be fully reckoned: Algeria remains a wounded society, so far unable to gain a sense of civic balance, while the ugly racism of the *colons* has been transferred to France itself.

Finally, the Algerian insurgency demonstrates, as have the American and Irish insurgencies, that there is a pattern or sequence that leads through actions that impose upon a people a sense of identity, to the use of terrorism to disrupt the administration of the occupying power, to almost inevitably successful guerrilla war, to, finally, the demise of the guerrillas. In this sequence, we should also see how small a part sheer military power plays. France had all the power and lost the war.

So it was with the French in Vietnam, to which I next turn.

CHAPTER 9

THE VIETNAMESE STRUGGLE AGAINST THE FRENCH

NO COUNTRY HAS EVER BEEN MORE STUDIED THAN VIETNAM. IT HAS been the subject of countless investigations, reports, and analyses, as well as hundreds of books and articles. However, the sheer volume of information has tended to overwhelm the meaning of what was being discovered. As I have mentioned in the introduction, when I was a member of the State Department's Policy Planning Council, I found that, despite the mass of material crossing my desk on Vietnam, no coherent analysis of the nature of its struggle in guerrilla warfare emerged. Nor did any consistent treatment of Vietnamese history. What American officials were trying to appreciate was not a country with a people who lived in a cultural tradition and had widely shared aims but a disembodied conflict between opposing military forces. At least in retrospect, it is clear that they were missing the "forest" by concentrating on the "trees." Here, of course, I cannot give a detailed history, but I will try to bring out what is essential to understand the nature of the guerrilla war that began against the French and carried on against the Americans. I begin with the land and its inhabitants.

As it exists today, Vietnam is a long sliver of territory, stretching roughly 1,000 miles south from the Chinese frontier to the Gulf of Thailand and the South China Sea. It varies in width from 300 miles near the Chinese frontier to less than 50 miles in the waist and fans out again in the south to about 180 miles. More important than the length is that the coast is broken by coves and inlets that give the country roughly 1,500 miles of seafront, so the Vietnamese have always been a maritime people. During the French occupation the majority of the people lived within a day's walk from the sea. Inland, the country is watered by two great rivers, the Red (Sông Hong), flowing south from China through Hanoi to the Gulf of Tonkin, and the Mekong, forming the border between Laos and Thailand and cutting through Cambodia to exit Vietnam in a delta west of Saigon. Between these river systems and the coast are mountain ranges, jungle, highlands, and low-lying rice fields astonishing in their variety for an area roughly the size of New Mexico.

Just as the land varies, so do its peoples. The Vietnamese describe their internal divisions in terms of climate and history: they joke that southerners, around Saigon, are lazy and slow-witted; those in the central area around Hué are hidebound and traditional; and northerners, around Hanoi, are aggressive and sharp-witted. As Douglas Pike found, "The Vietnamese are as conscious of region as an Indian is of caste . . ." In part, the sense of regionalism probably results from the way the Viet people arrived in their land. The earliest records placed them in southeastern China. From there, by fits and starts, they moved south, overwhelming the Champa kingdom, then centered in the area around the modern city of Hanoi in the eleventh century, and reaching the area around Hué in the early fourteenth century; by the end of the fifteenth century, after repelling a Chinese invasion, most settled in the central area. Some, however, continued to migrate south to take over the area around Saigon and the Mekong delta by 1780. The Viet and Champa peoples both hugged the seacoast and rarely penetrated far inland so that various non-Viet peoples, later collectively called by the French *Meo* (savages) or more politely *les montagnards* (the mountain people), continued to inhabit about four-fifths of Vietnam; they lived virtually undisturbed until modern times in the mountains and jungles of the hinterland.

Another group that was to play a significant part in Vietnamese history was the Chinese. China always overshadowed Vietnam. It was the source of much of Vietnamese culture and belief. Successive Vietnamese rulers feared it as the Asian superpower that might seek to incorporate their land and absorb their peoples as it had assimilated hundreds of other communities and Sinicized millions of aliens over its long history. To its neighbors, China seemed a great vortex, a sort of social "black hole" into which alien peoples poured, never to return. When, at the end of the Second World War, a Chinese army took over control of the north to effect the Japanese surrender, so alarmed was the Viet Minh leader Ho Chi Minh that he negotiated a five-year cease-fire with the French. When his colleagues berated him for this move, he said, "You fools! Don't you realize what it means if the Chinese stay? Don't you remember your history? The last time the Chinese came, they stayed one thousand years! The French are foreigners. They are weak. Colonialism is dying out . . . They may stay for a while, but they will have to go because the white man is finished in Asia. But if the Chinese stay now, they will never leave. As for me, I prefer to smell French shit for five years, rather than Chinese shit for the rest of my life."

It wasn't only in armies that the Chinese came. For centuries they had filtered into Vietnam as small groups of refugees. In the collapse of the Ming Dynasty in the seventeenth century, the first large group, some three thousand, arrived and were given asylum. By the end of the nineteenth century, others had come and the community had multiplied so that whole areas of most Vietnamese cities were Chinese.

Not long after the Viet people had established themselves in all of what was to become Vietnam, Europeans began to arrive. The first European who may have sailed along the Vietnamese coast, then the kingdom of Champa, was Marco Polo. He tells us that he was there about 1285 and mentions that its forests produced aloe and ebony and were filled with elephants. Whether it was he or his ghost-writer who wanted to catch the readers' attention, he also tells us that the king took to bed all the beautiful young women of the country and had produced 326 children. Unfortunately, after that "reader friendly" observation, he paid no further attention to the country.

Marco Polo was followed by a number of intrepid Franciscan and Jesuit missionaries. They were more interested in the commerce and capacity of the natives than in who took whom to bed so their accounts tend to focus on armies and agriculture. By the seventeenth century, Viet armies reached one hundred thousand men and elephants, the tanks of that period, five hundred. Few European armies could have then matched them. More important, the Viets produced goods that made profitable the trade of the early Portuguese and Spanish seamen in the Pacific. Picking up the scent on the trail blazed by the Portuguese and Spaniards, Frenchmen organized the Société des Missions Etrangères. As missionaries began to arrive seeking to impose an alien culture and religion, they found the local population so hostile that they often posed as merchants because commodities were more welcome than creed. When they openly proselytized, they were rounded up and deported, and subsequently many of their converts, whom the Vietnamese regarded as subversives, were imprisoned or killed.

Meanwhile, the Vietnamese were modernizing. Particularly in the northern (Trinh) kingdom in the early eighteenth century, the government carried out a series of administrative reforms that increased popular well-being. The collection of taxes was regularized, the penal code was made less brutal, the courts were rendered more honest, and help was given to open new agricultural lands. Fearing the encroachment of the Europeans, the Viets also reorganized their army on a system of conscription. Drawn from village communities in small numbers, this defense organization more resembled a guerrilla force than a European-style army. Each village was expected to contribute a contingent known as a *thuyen*, which was composed of around thirty men, but they were seldom used since, for many years, the country was at peace. Then, in the last years of the eighteenth century, the Nguyen kingdom of "the center" (what the French later termed Annam), based on Hué, embarked on the conquest of the north. After proclaiming himself emperor of all Vietnam, Nguyen Anh moved in 1803 to regularize relations with China on the traditional Chinese formula for a tolerated neighbor: he sent tribute and performed homage to the Chinese emperor. The dynasty thus established lasted until it was formally abolished by the Emperor Bao Dai and its mandate turned over to Ho Chi Minh in a ceremony in Hanoi on August 25, 1945.

What had made the Nguyen dynasty able to unify Vietnam was not just internal reform, better relations with China, or able rulers, but the arrival of the French. Their leader was the missionary Pierre-Joseph-Georges Pigneau, who came to Southeast Asia in 1765. After a series of escapes from hostile officials, he met the future emperor, Prince Nguyen Anh, who was then a fugitive. Pigneau became a sort of tutor to the young prince just as Edward Lansdale, two centuries later, would become the "tutor" first to the Philippine "prince" Ramón Magsaysay and later to the Vietnamese "prince" Ngo Dinh Diem. As Lansdale was to do with Magsaysay and Diem, Pigneau arranged the relationships of the prince with a foreign power. In 1784 he embarked with the son of Nguyen Anh, Prince Canh, on a trip that would take them to Versailles, where the prince cut a gaudy "Oriental" figure, met with Louis XVI, and (with the guidance of Pigneau) presented a plan for France to intervene in Vietnam. On November 28, 1787, just two years before the French Revolution, Canh concluded a treaty of alliance with France, in payment for which France was to receive strategically and commercially important territory adjacent to the imperial capital. Pigneau then came out from the shadow cast by the Church to become the first French commissioner for Vietnam.

Recognizing the weakness of the French government, which was already tottering toward the Revolution, Pigneau set out to mobilize Frenchmen to do privately what the French government could not afford to do. He had used his time in Paris with the attractive young prince to convince a number of aristocrats and merchants to contribute to the French venture he wanted to organize. With the money he collected, he recruited a force of several hundred mercenaries. When they reached Vietnam, Nguyen Anh had already managed to recapture Saigon. A confused period followed in which the members of the dynasty and their rivals jockeyed for power, but with additional soldiers recruited by Pigneau and French "advisers" brought in to help organize the navy and train the army, Nguyen Anh slowly recaptured the country. By 1803, he had it all.

Thereafter, both the French and the Vietnamese pulled back from their alliance. Elsewhere engaged, Napoleon lost interest in the Far East. It was simply beyond his field of vision; India, on which he had his eye, had already proved to be a country too far. Then, as I described in Chapter Two, Napoleon's army became bogged down in a guerrilla war in Spain. It was not until after Napoleon's defeat and fall that in 1817 French merchants began to sail from Bordeaux to Vietnam to buy its silks and exotic woods and French missionaries again carried their message eastward.

As the pace of Western intrusion picked up, the successors of Emperor Nguyen Anh began to worry about its effect on their people. The dynasty then also felt secure and no longer needed the French. Prince Canh, who had been entertained as a youth at Versailles, had aged and died. His brother, who became emperor, had not partaken of the delights of the French court and saw the French in a rather different light. A devoted Confucian, the Emperor Minh-Mang viewed the French missionaries as subversive agents out to destroy Vietnamese culture and religion. He wanted them and all their local allies out of his country, so he broke off diplomatic and even commercial relations with France and began to persecute the Vietnamese who had converted to Christianity.

His actions would have made little difference in the eighteenth century, but in the early decades of the nineteenth century Europeans were growing in power relative to the Asians and were becoming more aggressive. The British, intermittent allies of the French, were then pushing into an increasingly decrepit China. In 1841 the British established themselves at Hong Kong and forced the Chinese to open five of their ports to Western commerce and resident Europeans. Both encouraged by the British and competing against them, the French also took a more aggressive stance. Even the Americans tagged along. The Vietnamese could have bowed to the new winds blowing across Asia from Europe, but they chose to resist. To them that meant trying even harder to prevent subversion by missionaries. Wherever they could, they arrested and imprisoned them.

Goaded by religious zeal and adopting a role as *"missionnaires armés de la civilization européenne,"* the French moved against the recalcitrant Viets: in 1843 a French naval vessel threatened to bombard the port just south of the imperial capital at Hué if imprisoned missionaries were not released. Threats were quickly followed by cannon fire. (Eager to show its new flag, the American frigate *Constitution* also bombarded a Vietnamese port.) Gunboat diplomacy had arrived. Under Napoleon III, France was looking for an excuse to move into the country, and the Vietnamese provided what France wanted: it rejected a French ultimatum. The result was that in 1858, no longer satisfied by bombardments from the sea, a French infantry force went ashore for the first time. It landed at Tourane (Danang) just south of the capital at Hué. The next year, the French moved south to take Saigon. They were then temporarily diverted by military operations against China, but in 1861 they landed a much larger force that conquered the entire south, what they called Cochin—a corruption of the Vietnamese word *ke chime.* In the peace settlement that followed, most of the south was ceded to France and became the nucleus of a new colony. Beginning there, France was to remain in Indochina for nearly a century.

What the French found when they arrived was a fully articulated society based on kinship interwoven with Confucianism, Buddhism, and Taoism. The basic social unit was the household, known in Vietnamese as the *nha* or *gia.* It was the household, rather than the individual, that was responsible for taxes and that was counted in the census; its importance in society was reinforced by the Confucian emphasis on filial piety. The household was a place where ancestors were honored, and their graves incorporated family history into the buildings. Households gathered into villages that were more than mere collections: they acquired a corporate identity that was expressed in a variety of ways. Each household paid to the village a tax, usually in grain, that enabled the village to sustain itself in periods of drought or flood; the village as a corporate unit could own property, had its own elected officials, and accepted responsibility for public order and defense of the households. Perhaps as important, the villagers participated in a variety of associations that laced together the inhabitants by trade, crafts, religion, amusement, and mutual assistance ranging from birth to the grave. The village was more than the "nation" writ small: for the inhabitants it *was* the nation. We will later see how significant this all-encompassing system was when the Americans uprooted the inhabitants and tried to replace their villages with "strategic hamlets."

This focus on local loyalty contributed to clandestine organization. Villages were small and weak—we have seen that their "armed forces" usually amounted to only thirty or so men—and the easiest way to protect themselves was to carry on many of their activities in secret. In nearly a thousand years of resistance against the Chinese and during the turbulence of the civil wars in which their various rulers struggled to control their bodies and assets, they learned to lead what amounted to an underground life. This would play a significant part in the ease with which the secret societies arose. In their turn, the Communists were able to operate under the eyes of the French, who struggled for nearly a century to cope with a political landscape that they could only partly see. As one French officer put it, "We . . . walk in a hostile country as though blind." Americans would later suffer the same disability.

Buddhist monks led villagers in small-scale guerrilla attacks on French soldiers from sanctuaries they set up in such virtually inaccessible places as the swampy Mekong delta. As the French commander wrote in 1862, "We have had enormous difficulties in enforcing our authority . . . Rebel bands disturb the country everywhere. They appear from nowhere in large numbers, destroy everything and then disappear into nowhere." As one area was suppressed, resistance was organized elsewhere; thus, ironically, it was in part the Vietnamese attempt to protect themselves in the south (Cochin) that caused the French to attack the center (Annam) and the north (Tonkin) and to turn them into protectorates.

To solidify their rule, the French set out to win over what remained of the Vietnamese bureaucracy, the mandarins. As the man who was perhaps the most effective French governor-general, Jean-Marie de Lanessan, explained his policy, it was based on the notion that "in every society there exists a ruling class, born to rule, without which nothing can be done. Enlist that class in our interests." The mandarins were initially willing to collaborate, but they soon realized that collaboration was leading to the replacement of their way of life by an alien system and their religion by Catholicism; so instead of "pacifying" their fellow countrymen, they began in the last years of the nineteenth century to lead them against the French. The most famous man of this class, who had served as imperial censor and is still a hero among the Vietnamese, Phan Dinh Phung, created a model for future insurgents. Having rallied a guerrilla force of some three thousand men, he created a small guerrilla state in the hinterland of central Vietnam that supported itself by collecting taxes, administered its area by recruiting Vietnamese officials, and even set up workshops to manufacture guns and ammunition. Alarmed by Phan Dinh Phung's challenge, the French began to experiment with counterinsurgency.

Counterinsurgency begins with an attitude toward the insurgents. As we have seen, most regimes refer to them as "bandits." The common word among the French for the Viet insurgents was "pirate." The designation itself seemed to justify Draconian reaction. Even the more enlightened French officials did not oppose brutal suppression, but at least some already recognized what Mao Tse-tung would later highlight in his work on guerrilla warfare, *Yu Chi Chan,* that the insurgent, even if he were a pirate or bandit, could not pose a threat or even exist without the support of the people. So, as the influential French administrator Louis Lyautey wrote, "the pirate is a plant which will grow only in certain soils, and that the surest method [to defeat him] is to make the soil uncongenial to him." That was the first step. Then, having turned the people away from the insurgent, the French sought to turn the natives from "passive instruments" into "intelligent and voluntary collaborators." Lyautey's favored tactic was what his mentor Joseph Gallieni had called *tache d'huile*, an oil spot that would spread area by area over the countryside in "a methodical, necessarily slow expansion of French control." Gallieni and Lyautey were precursors of Robert Thompson and Edward Lansdale in their search for sophisticated counterinsurgency in the later American campaign in Vietnam. The problem was how to do it.

Perhaps the shrewdest counterinsurgency tactic the French adopted was designed to show that the rebels could not defend what was most sacred to them, the remains of their ancestors. In the ultimate humiliation, the French dug up the graves of Phan Dinh Phung's ancestors and put the bones on display in the market of a town they had "pacified." Thus, in the context of Confucianism, what they were proclaiming was that the "Mandate of Heaven," the divinely ordained right to rule, had passed from him and his partisans. In their failure to protect even the resting places of their ancestors, the Vietnamese nationalists showed their weakness, while in destroying them, the French showed their power and (they thought) proclaimed their legitimacy. This psychological assault may be compared to what the British were later to do against the Mau Mau in Kenya: they were probing into the deeper recesses of men's beliefs. Probably that, as much as their modern arms and military discipline, was how the French wore down the guerrillas. Even so, overcoming resistance in the lowlands took the French ten years, and fifteen in the more remote highlands. They bribed, offered compromises, and used starvation; the search and destroy tactics they employed were brutal. As Stanley Karnow remarked, "They gave no quarter. One of the French generals, refusing to regard his Chinese and Vietnamese foes as regulars, beheaded all his prisoners as 'rebels.'"

To give him his due, General Gallieni did not favor such a brutal policy of pacification. He was certainly not above using socially destructive tactics, as he did by fomenting among the non-Viet peoples, the Thais, Thos, and Muongs, ethnic "hatred and rivalries [against the Viets] which we must use to our profit," but he also favored economic and administrative assistance, even rebuilding what the soldiers had destroyed, immediately after one of his oil spots had been secured. It was largely due to Gallieni that the first phase of the Vietnamese war of national liberation, which began in 1858, wore down in defeat finally in 1898.

As they did in Algeria and were to do in the Middle East, the French used culture as a weapon of imperialism: to the extent possible the Viet language was replaced by French. Even primary schools were to be conducted in French; when it turned out that they could not furnish enough teachers, the administrators simply closed the schools. "Consequently, after ten years of foreign rule, a formerly literate region with more than 1.5 million inhabitants had fewer than 5,000 primary school pupils. Higher education had ceased altogether." The French would allow no Vietnamese efforts to fill the void since the "French regarded the efforts of the 'natives' to educate themselves along lines chosen by themselves as a political act against the colonial regime." As the *colons* advised the incoming French governor-general in 1886, "an educated native no longer meant just 'one coolie less' but one rebel more." In alarm, the French closed the most prominent native school and arrested its teachers.

Then in the first years of the twentieth century, all Asians were stunned by the 1905 Japanese defeat of the Russians; it was the first time that an Asian nation overthrew a major Western power. It started a process that would change the way the Vietnamese saw the French. Suddenly Japan seemed a model not only for disaffected students and their teachers but even for members of the royal family. As Joseph Buttinger, perhaps the most sensitive scholar on this period, has written, "A period of intense educational activity accompanied this awakening of the national spirit under the banner of 'modernization.' Study groups formed, patriotic merchants supported the publication of small newspapers that proclaimed eventual freedom for a Westernized Vietnam, traveling lecturers assured their listeners that education was a means of liberation and that Vietnam would rise again if it followed the path chosen by Japan."

The French responded, as usual, with main force, but that policy could not be applied uniformly. France wanted bodies and could not separate them from minds. In the First World War, the French desperately needed both workers and soldiers and took them from all over their empire. They included about 100,000 Vietnamese who were forcibly recruited and sent to France. There they not only mingled with the French on something approaching equality but also began to absorb French ideas. Just as in Algeria, from which, as I pointed out in Chapter Eight, 173,000 Algerians were drawn into the French army, so in Vietnam it was the French spur that dug into the colonial flank. Like the Algerian Messali Hadj, so the young socialist Nguyen Ai Quoc, who would become known as Ho Chi Minh, joined French political movements that opposed imperialism and stayed on after the war in France. Messali Hadj and Ai Quoc were obviously exceptional men, but the contrast between the way they were treated in France and what they encountered on their return to the colonies affected thousands of others: it intensified rather than diminished rebellion against the colonial administration. A similar experience in Spain and Germany, as I pointed out in Chapter Three, had energized the first generation of Filipino nationalists. The French reacted as the Spaniards had done: rather than easing some of the galling restrictions, they tightened them. In this policy they were pushed by the growing *colon* community, which was utterly opposed to any moves toward liberalization or integration; its aim was to keep Indochina as a source of cheap labor producing raw materials for France's industry. French officials saw no need to compromise. By 1930 France had recovered sufficiently from the First World War to muster a powerful military force in Indochina. Thereafter, France's military fist was never gloved. In almost continuous search and destroy campaigns about 10,000 Vietnamese were killed, 50,000 deported, and upward of 10,000 taken off to the concentration camp on Pulo Condore (Con Son) island.

To rule Vietnam, the French relied heavily on three overlapping groups, the Catholics, the wealthy landlords, and both Viet and ethnic minority local troops. The Catholic minority aided in administration and generally knew what was happening in the cities; the wealthy had been given their lands when the administration proclaimed that the peasants did not own it, but that ownership was vested in the state. So, to win over the new upper (and collaborating) class, it gave its supporters land, thus necessarily creating a very large class of landless laborers. The French thought their policy had paid off when, with the help of the new landlords, they easily beat down a revolt in 1930. In the short term, they were right. For ten years a deceptive peace reigned.

Then, in 1940, following the French collapse in Europe, the Japanese took over the country. Until the last days of the war, they were content to rule through the French, but they forced upon the French increasingly unpopular policies while weakening the ability of the French to enforce them.

The group that was poised to take advantage of this new situation was the central committee of the Vietnamese Communist Party (the Lao Dong). It had lived in the shadows during the previous decade and so had remained at liberty. It decided in May 1941 to form a front, comparable to the EAM in wartime Greece, to gather together all the nationalists into a single anti-French front. The Vietnamese Communists called their front the Viet Nam Doc Lap Dong Minh Hoi (League for the Independence of Vietnam), or Viet Minh.

The Communist leaders of the front wisely defined their program to fit Vietnamese conditions rather than Marxist theory. Instead of basing it on virtually nonexistent industrial workers, as the Marxian model suggested, they targeted the peasantry. The core of their policy objectives were nationalism (getting rid of the foreigners, both the French and the Japanese) and justice for the exploited peasantry (land distribution). These were the two issues that captured the attention and loyalty of more Vietnamese than any other. The Viet Minh program theoretically meshed well with the stated objectives of the Allies and the general sentiment of their peoples.

In America, the leaders of both parties expressed their determination to end imperialism. The Republican presidential candidate Wendell Willkie stressed that "this war must mean an end to the empire of nations over other nations," and President Roosevelt responded that this was a "well-accepted" policy. Roosevelt, presaging future moves by the as-yet unformed United Nations, began the planning of an international force to replace foreign armies of occupation.

Meanwhile, in the field, the Allies had practical reasons to oppose the French. The French in Vietnam were under the control of the collaborationist Vichy government and were working closely with the Japanese. The Americans needed intelligence to better fight the Japanese in the South China Sea and Burma and to rescue the crews of aircraft shot down over Indochina. The only effective Vietnamese group the Allies identified was the Viet Minh. So it was the group the Americans contacted. One team from the OSS, the wartime precursor of the CIA, parachuted into the North Vietnamese mountains, where they encountered a desperately ill Ho Chi Minh. An American medic probably saved his life. What they saw of his group did not encourage them: it was composed of only about two hundred guerrillas armed with what was more a museum of antique muskets than an arsenal.

Always, the quest for arms could shape the guerrillas' tactics. As Vo Nguyen Giap, the future victor of Dien Bien Phu, told Stanley Karnow nearly half a century later, he began operations with a team of only thirty-four men and women who used their antique arms to overpower first one small French outpost and then another on Christmas eve of 1944. These engagements netted them more modern arms and the psychological boost of success. Beginning there, Giap performed remarkable feats of organization. As they grew in number and spread out over Vietnam, the Viet Minh became "the sole efficient resistance apparatus within Vietnam" and so became valuable to the Allied forces. By early 1945, Giap's guerrillas had grown to about five thousand. Consequently, as in Yugoslavia and Greece, the Allies parachuted teams and equipment into areas the guerrillas controlled. In the Balkans, as I have described in Chapters

Five and Six, they were mainly British, while in Indochina they were mainly American. Impelled by wartime strategy, the Americans literally jumped in with what the *Pentagon Papers* described as "modest" aid, thousands of modern weapons and tons of supplies, and began training the guerrillas on how to use them.

As the fighting went on, the Allies were setting a lofty tone for the postwar world: freedom was the word that figured in every statement. The words were stirring, but what really changed the thought of the Vietnamese was, ironically, an action by the Japanese. The Japanese were content to allow the French to administer Vietnam under their overall control, and the French were willing to do anything the Japanese demanded, even to force the peasants to give up so much of their rice as to cause widespread famine in which perhaps half a million Vietnamese starved. The French façade served Japanese purposes so economically that Vietnam was the only country in Asia where they tolerated a colonial regime. They even allowed it to retain a hundred thousand soldiers, which was about three times the size of the average contingent of Japanese forces. Then, in March 1945, sensing that the French were about to switch from Vichy to the Free French, the Japanese struck. Within hours, they dissolved the French administration, disarmed French troops, humiliated or beat Frenchmen in the streets, and raped their women. These actions destroyed the image of the French as the master race as completely as the American Indians had destroyed that of the British in the massacre of Braddock's redcoats in the French and Indian war.

The Japanese did not stay on top for long. As the war wound down, it was clear that their days were numbered. Recognizing this, Ho Chi Minh bombarded the Allied leaders with what are called in the American government documents "eloquent appeals for U.S. or U.N. intervention in Vietnam on the grounds of the principles embodied in the Atlantic Charter, the U.N. Charter, and on humanitarian grounds." *None of his messages ever received a reply* and all were secreted away for the next quarter century. Ho refused to be discouraged, and realizing that some form of transitional oversight would be demanded by the Allies, he informed one of the OSS men assigned to him that he would welcome "a million American soldiers . . . but no French." However suspicious they were of the Vietnamese, the Americans did not then favor a return of the French.

Even before President Roosevelt's death, however, there was a major shift in policy toward supporting the return of the French to Vietnam. It accelerated after his death. Urged by General Charles de Gaulle, Secretary of State Edward Stettinius "told the French representatives [at the San Francisco conference that founded the United Nations] that the United States had never 'even by implication' questioned French sovereignty in Indo-China . . ." A few months later, Acting Secretary of State Dean Acheson declared "that the United States had 'no thought of opposing the reestablishment of French control.'" It was this policy that made the United States a party to the war.

The shift in the American position on Vietnam was largely determined by anticipation of the cold war. As anticipation turned to reality, fear of what came to be called the "domino effect" emerged. As put forth, it was that "French resistance to Ho Chi Minh . . . [was] the crucial link in the containment of communism." Otherwise, state after state would fall to Communism as General Smuts had warned Churchill when discussing Greece. That was not the belief—or hope—of the French Communist Party, which took the same position on Vietnam that it took on Algeria: its representatives voted in the National Assembly for increased funds for the French force that would fight the Viet Minh and generally staked out a position more nationalistic than the French right.

Ho paid no attention to what either the Americans or the French Communists said. He focused on strictly local issues, using the confusion of the collapse of the Japanese to strike a deal with the only recognized local government, that of Emperor Bao Dai. Correctly judging the fragility of his position, Bao Dai asked the Viet Minh to form a new national government, and Ho, recognizing the reality of his power, insisted that the emperor abdicate so that Vietnam could become a republic. In a grand ceremony on August 25, 1945, Bao Dai turned over the insignia of rule to a delegate of the "Democratic Republic of Viet Nam." A week later, Ho arrived in Hanoi to proclaim the new era in a speech patterned on and drawing phrases from the American Declaration of Independence. He sought to portray the Viet Minh struggle as more against the Japanese, America's enemy, than against France, America's ally. But events were taking quite a different turn.

That turn involved the Allies flooding north Vietnam with 150,000 Chinese Nationalist troops and south Vietnam with a smaller but more active British force. Their stated objective was to disarm the Japanese. The commanders of the two forces saw their roles differently: the Chinese used their opportunity to pillage—and to corner the opium crop—while the British, aided by Japanese troops, restored the French administration. As quickly as American ships could be organized, the French army returned. On October 25 the French armored division that had been allowed to liberate Paris came ashore. The French premier announced on December 23, 1946, that "before all, order must be reestablished, peaceful order which is necessarily the basis for the execution of contracts." His putting "security" before sovereignty was the classic mistake of external governments; that sequence has rarely worked and did not work then or later in Vietnam. Thus, events of the following weeks were to set the tone for the following years: heavily armed with American equipment, the French army moved through the countryside like a paddle through water. It pushed aside all opposition, but as it moved the "water" flowed back into position. The French would never again be able to establish "peaceful order." They had long since forgotten the wise words of their nineteenth-century commander, General Gallieni, that "a country is not conquered and pacified after military operations have bent all heads by terror; when the first fear is overcome, ferment of revolt will grow among the masses, which the accumulated bitterness caused by brutal action of force will multiply and steadily increase." It did. And soon.

As John McAlister summarized months of military actions and forecast the years of combat that followed:

Once the French forces were dispersed over the expanse of southern Vietnam, they found themselves vulnerable to guerrilla attack . . . Thus the French came quickly to confront what was to be the main problem of the guerrilla phase of the Indochina War: how to divide their forces between those assigned to static defense elements and those given mobile intervention duty. This is perhaps the classic guerrilla war problem for the forces attempting to establish or maintain government authority. If forces are concentrated in order to wipe out an inferior guerrilla band, the adversary merely refuses combat and takes the occasion to hit emplacements left unprotected by the concentration of government forces. If government forces are dispersed to provide static security for routes of communications, military depots, economic installations, and a scattered rural population, then guerrilla forces concentrate to a strength sufficient to overpower the defenders [of a single installation] and disrupt communications or capture supplies.

Such tactics, of course, were only the military aspect of the Viet Minh strategy. The part on which the Viet Minh focused most of its attention was first political and then administrative. Politically, Ho was already the only recognized national leader, and every effort was made to keep him before the Vietnamese public and to contrast his actions with those of the almost always abusive and corrupt Vietnamese who worked for the French. But administration turned out to be the Viet Minh's most impressive action. They enrolled first tens and ultimately hundreds of thousands of Vietnamese to fight for or support their combatants, controlling the whole operation with only about five thousand party members. They organized, converted, or rebuilt factories to manufacture arms so that within a year they were turning out thousands of pistols, rifles, and even machine guns. And, most important, they sent their cadres of party workers into the void the French had created in the villages when they overturned what Paul Mus called the "Confucian balance" that had regulated the lives of the inhabitants. Just as the EAM did in Greece, the Viet Minh encouraged the villagers to form committees that spanned every conceivable activity so that virtually every villager became a member of at least one. It was this combination of politics and administration that enabled the Viet Minh to organize for the next phase of the war with France.

That phase started badly for the Viet Minh. Like most military leaders, General Vo Nguyen Giap wanted, as Mao Tse-tung had done, to replace the highly successful guerrilla forces with a regular army. The French were delighted. In formal warfare, they had the advantage. As one French general said, "If those gooks want a fight, they'll get it." France rushed in troops and their best generals. When he arrived in December 1950, Jean de Lattre de Tassigny publicly proclaimed that he could "win a decisive victory within fifteen months." Privately, he was sure he could not. His successor Henri Navarre said, "A year ago none of us could see victory. There wasn't a prayer. Now we can see it clearly—like light at the end of a tunnel." His phrase would echo down the years in Paris and Washington, but then Navarre was partly right: Giap had moved toward conventional warfare too quickly. He carried out three premature attacks, and the French hit his forces devastating blows.

So elated were the French by their military successes that they decided to lure the Viet Minh into what they thought would be the killing field of Dien Bien Phu. They were too late. Giap had learned from his mistakes, had built his forces to some fifty thousand front-line soldiers backed up by reserves of another twenty thousand, and had laid down a remarkable military infrastructure: he was ready for them. The French had only thirteen thousand men at Dien Bien Phu, and only half of those were combat-trained soldiers. After a grueling siege, in which they were starved and worn down, they surrendered.

Dien Bien Phu was the graveyard of the French Empire in Indochina. In the next chapter, I will show how America came into that graveyard.

CHAPTER 10

AMERICA TAKES OVER FROM FRANCE IN VIETNAM

THE FIRST AMERICAN MILITARY INTERVENTION IN VIETNAM TOOK place a century before the OSS made contact with Ho Chi Minh during the Second World War. It happened on May 20, 1845, when a group of marines landed from the USS *Constitution* near the imperial capital Hué to try to rescue a French missionary who had been imprisoned for what the Vietnamese emperor regarded as subversion. The marines were rebuffed, got back on "Old Ironsides," and sailed away. Their failed foray was the first example of what some of their great-grandchildren would be accused of, "cutting and running."

Their "run" was a long one. The U.S. government had no further interest in Vietnam for nearly a century. Then it tried to shield Vietnam from the horrors of the Second World War. Shortly before the Japanese attack on Pearl Harbor, it urged the Japanese government to withdraw and offered in return to treat Vietnam as neutral territory. The Japanese refused. Their invasion made Vietnam part of the worldwide struggle.

Sustained American intervention in Vietnamese affairs began

during the Second World War. What America sought, however, was ambivalent: on the one hand, President Roosevelt promulgated the Atlantic Charter, which called for national self-determination. He told the British ambassador in Washington that Indochina "should not go back to France but that it should be administered by an international trusteeship." Roosevelt got both Chiang Kai-shek and Josef Stalin to agree to his idea of trusteeship, but Churchill, fearing the precedent that this might set for India, was adamantly opposed. So the idea was dropped. But Roosevelt was appalled by the French record of colonial administration, writing that "France has had the country—thirty million inhabitants for nearly one hundred years, and the people are worse off than they were at the beginning . . . France has milked it for one hundred years. The people of Indo-China are entitled to something better than that." However, attempting to motivate the French to join the Allied side, Roosevelt repeatedly expressed his support for the reestablishment of the French empire "throughout the territory, metropolitan or colonial, over which flew the French flag in 1939."

What clarified and solidified the wavering American position on French rule of Indochina was the growing perception of a threat of Communism as a weapon of the Soviet Union. Before his death, Roosevelt had begun to shift his thinking away from the idealism of the Atlantic Charter, but it was President Harry Truman who pushed that trend forward to the cold war. Unequivocally, Truman assured the French that the American government "never questioned, 'even by implication, French sovereignty over Indochina' [and it would] expect France to decide when its peoples would be ready for independence."

But American support for France was not unqualified. To put pressure on the French "to decide" on independence, American ships were ordered on January 15, 1946, not to carry French troops or war matériel to Vietnam. It was the British who enabled the French to bring their army to Vietnam (in American-supplied ships) and supplied them with American equipment.

As the *Pentagon Papers* indicate, "U.S. policy toward Vietnam then shifted . . . to the area of the U.S. relationship with France." In the growing determination to unify Western Europe against the USSR—a series of moves that led toward the "European Defense Community" (which France ultimately did not join)—Vietnam's aspirations for freedom were shunted aside. Recognizing the growth of French power and worried by the intrusion of the Chinese in the north, Ho Chi Minh, as president of the newly proclaimed Democratic Republic of Vietnam, signed a cease-fire on March 6, 1946, "to welcome amicably the French Army . . ." This was a victory for French diplomacy, but the French high commissioner was not pleased: he "recorded his 'amazement that France has such a fine expeditionary corps in Indochina and yet its leaders prefer to negotiate rather than to fight . . .'" His policy prevailed; as the authors of the *Pentagon Papers* write, "The record shows that through 1953, the French pursued a policy which was based on military victory and excluded meaningful negotiations . . ."

Could the French prevail? The American government was warned that they could not. John Carter Vincent, whose career was subsequently ruined by hostile reaction to his insights on Vietnam and China, was then director of the Office of Far East Affairs in the State Department. On December 23, 1946, he presciently wrote the secretary of state that "with inadequate forces, with public opinion sharply at odds, with a government rendered largely ineffective through internal division, the French have tried to accomplish in Indochina what a strong and united Britain has found it unwise to attempt in Burma. Given the present elements in the situation, guerrilla warfare may continue indefinitely." This chapter will illustrate how it did.

Despite the warning, growing American fear of Communism led to identification of the United States with France. On May 13, 1947, American overseas posts were telegraphed, "we cannot conceive [of] setbacks to long-range French interests which would not also be setbacks [to] our own . . . Following relaxation [of] European controls, internal racial, religious, and national differences could plunge [the] new nations into violent discord, or already apparent anti-Western

Pan-Asiatic tendencies could become [the] dominant policy force, or Communists could capture control." In contemporary Greece, General Smuts had already foreseen a similar "wave of disorder and wholesale communism" sweeping over Europe. The image of falling "dominos" was to haunt Western statesmen for a generation.

Then, apparently suddenly, senior officials in Washington began to ask one another about the nature of the Viet Minh movement. Although Ho Chi Minh was believed to be a Communist, there was "the seeming paradox" that "he seems to have no visible ties with Moscow . . . Ho seems quite capable of retaining and even strengthening his grip on Indochina with no outside assistance . . ." Was it possible that he was an Asian Tito? What gradually forced these speculations into a firm attitude was the Greek insurgency that brought about the Truman Doctrine. America saw the Soviet hand everywhere. For Secretary Acheson, the "recognition by the Kremlin of Ho Chi Minh's communist movement . . . should remove any illusions as to the 'nationalist' nature of Ho Chi Minh's aims and reveals Ho in his true colors as the mortal enemy of native independence in Indochina." The day after Acheson spoke, the United States recognized the quasi-independent state that France proclaimed under Emperor Bao Dai.

Was the decisive rejection of Ho inevitable? The authors of the *Pentagon Papers* comment that "U.S. insistence on Ho's being a doctrinaire communist may have been a self fulfilling prophesy . . . the U.S. offered Ho only narrow options." As I have recounted, he received no replies to his frequent appeals [for American recognition and even military occupation]. As they summarize, "The simple truth seems to be that the U.S. knew little of what was transpiring in Vietnam, and certainly cared less about Vietnam than about France. Knowing little and caring less meant that real problems and variety of choices were perceived but dimly . . . President Eisenhower's later remark about Ho's winning a free election in Vietnam with an 80% vote shone through the darkness of our vision about Vietnam; but U.S. policy remained unillumined."

Two events in Asia were decisive in forming American policy

toward Vietnam: the first was the outbreak of the Korean War (June 25, 1950—ended in a cease-fire on July 27, 1953) and the second was the Viet Minh defeat of the French at Dien Bien Phu (March–May 1954).

The North Korean invasion of South Korea with the subsequent Chinese overt and Soviet covert intervention convinced the American government that America was engaged everywhere in mortal combat with Communism. Already in February 1952, President Truman had approved a National Security Council finding that the United States might be forced to take military action even in Vietnam. Consequently, when the French forces at Dien Bien Phu were on the brink of defeat, the United States allegedly (because if there were documents they have disappeared) considered a massive air strike—code named "Operation Vulture"—on the attacking Viet Minh troops. Reportedly there was a French attempt to get the United States to use a nuclear weapon. However, the United States took no action. The French surrendered, and the war ended in the Geneva Accords that aimed to temporarily divide Vietnam on the 17th parallel.

To the United States, the Geneva Accords were anathema. The Eisenhower Administration refused to sign them but reluctantly agreed to abide by them. However, it determined to keep French power effective in Vietnam. By 1954 it was financing 78 percent of the costs of the French military establishment there, $1.063 billion. Secretary of State John Foster Dulles then tried to create a ten-nation commitment to military intervention, but got only Thailand and the Philippines to agree. On that basis, the Congress would not approve action, and without congressional approval, Eisenhower would not act. However, the U.S. Army worked out a plan for intervention; it estimated that twelve U.S. divisions would be required to conquer and "regime change" Vietnam if the Chinese intervened and seven if the Chinese did not. The army pointed out, however, that such a venture would "seriously" affect America's ability to meet its NATO obligations. So America began to develop the alternative strategy that it would follow for nearly twenty years.

The key element in this strategy was the preservation of South

Vietnam. Ironically, it was largely because of the Chinese Communist representative, Chou En-lai, that this option became available: at Geneva he encouraged the concept of two Vietnams, reaffirming traditional Chinese policy to keep Vietnam weak. A divided Vietnam with the South controlled by the French seemed to him the best alternative available. His stand provoked the fury of the Viet Minh delegate, who muttered to an aide, "he has double-crossed us." The United States position was roughly parallel to the Chinese: keep Vietnam divided, with the "free part," the South, kept viable with or without French assistance.

Keeping the South viable required that the elections, to be held in July 1955 as called for in the Geneva Accords, not be held, since America assumed that the Viet Minh would win and would then take over the entire country. When the agreed time to hold them came, the government of the South (with American support) refused. Avoiding them was partly justified because of the massive exchange of populations between the two parts: with U.S. overt logistical support and as a result of U.S. covert fear-mongering, nine hundred thousand civilians moved from North to South. Since a significant portion of them were Catholic and virtually all were strongly anti–Viet Minh, they would play a major role in Diem's control of the South. At the same time, more than one hundred thousand Viet Minh militants and civilians favorable to them moved north. Both sides used the transfer to create "stay-behind" units that would covertly carry on the fight.

The contrasting fate of the two stay-behind groups demonstrates conclusively the importance of Mao Tse-tung's concept of the "sea" supporting the "fish." The anti–Viet Minh "stay behinds" who were encouraged by the CIA to remain in the North to gather intelligence and engage in sabotage lacked a popular base; so, unable to get supplies, shelter, or information from the Northern civilians, they were quickly caught and presumably killed. When they disappeared, the CIA attempted over a number of years to introduce new groups. The attempt was described by the then Saigon station chief of the CIA (and its later director), William Colby, as "notoriously unsuccessful"—all the teams parachuted or landed from ships quickly stopped radio communication or were "'doubled' by their Hanoi captors and were sending messages designed to lure more teams to capture or death." In the North, these "fish" found no "sea" in which to swim. In contrast, the five thousand or so Viet Minh "cadres" who stayed in the South survived because they were protected by sympathizers. Clearly, as Vo Nguyen Giap boasts, the Viet Minh "military and political struggles are closely coordinated, assist each other, and encourage each other to develop. This coordination is a law of the revolutionary struggle in our country." It worked for the Viet Minh, who were after all Vietnamese, but not for the Americans, who after all were foreigners.

The man who conceived the CIA "stay-behind" program for the North, Edward Lansdale, figured prominently in my account of the Philippines as the man who taught Ramón Magsaysay how to fight the Hukbalahap guerrillas and, indeed, how to be president. I suggested that Lansdale aspired to be a modern Machiavelli with Magsaysay as his "prince." So astonishing did his success there seem to the men in Washington that the CIA decided to send him to Saigon "amid the despair after the fall of Dien Bien Phu." When he asked what precisely he was to do, Secretary of State John Foster Dulles replied, "Do what you did in the Philippines." He found the clay from which to mold another prince in Ngo Dinh Diem.

Neither the French nor the Americans thought Diem was good clay, but the choice was limited. So, having picked him, they had to help him build a base. The CIA arranged this by maneuvering Emperor Bao Dai into appointing him prime minister. He quickly ousted the emperor and made himself president. That was a necessary first step, but Diem had to be taught how to perform his new role. He had to be tutored in the same way as Magsaysay. To do that, Lansdale virtually moved in with Diem, as he actually did with Magsaysay, so that he could be constantly at hand when decisions were made or options were evaluated. Machiavelli never had such a relationship with his prince, Lorenzo Medici. "Diem might have been Washington's choice in Saigon," wrote Neil Sheehan in his valuable study of counterinsurgency in Vietnam, "but he could not have survived without

Lansdale at his side . . . Lansdale was at the palace almost every day and spent many of the nights with Diem—encouraging him, planning their moves, calling the plays with tactical expertise that he had learned in the Huk war and that Diem lacked. Without Lansdale's guile, his intuition for the bold stroke, and the reputation he had acquired with the powerful in Washington because of the apparent miracle in the Philippines, Diem would have been swept away . . . Diem had no following beyond the Catholics . . . South Vietnam, it can truly be said, was the creation of Edward Lansdale."

Diem listened to Lansdale, but he had his own agenda. It did not include building a broad administration. Even with his Catholic supporters, Diem refused to share power. Rather, he set out like the Greek dictator Metaxas to create a one-man despotism. He played his thirteen separate security services against one another and appointed to military commands incompetent sycophants who were unlikely to gain enough strength to pose a threat to him regardless of what they did or did not do to the Viet Minh. Also like Metaxas, he created a fanatical youth movement patterned on Mussolini's Black Shirts. Diem's youth wore green shirts, but like the Black Shirts used the fascist salute.

The role of France in Vietnamese affairs had already been diminished by American action and inaction. France was reluctantly beginning to follow the American lead. At a conference in Washington in September 1954, despite its belief that Ngo Dinh Diem would be unable to unify or stabilize the country, France hesitantly agreed to support him. Partly to further reduce French influence, Secretary of State Dulles created another of the regional pacts that typified his approach to world affairs, the Southeast Asian Treaty Organization (SEATO). Then he moved to "cut out the French as middle-men in all its assistance for Vietnam, and began to deal directly with Diem, his government, and his armed force." After considerable "arm twisting," the U.S. Joint Chiefs of Staff agreed to supply a training mission for the troops of South Vietnam provided it had "safeguards against French interference." That first mission arrived in November 1954.

For the next two years, American officials complained of Diem's petty tyranny and corruption, but they felt that the Diem regime was at least temporarily necessary. Aware of their desire to replace him if they could find a viable alternative, Diem moved to smash his only non–Viet Minh rivals. So, seeing that they had no choice, his American critics swung to his support. But the period of apparent stability did not last long. His increasingly repressive regime was a virtual factory for opponents. The most competent and honest among the officer corps and the urban intellectuals were the main victims. Of these, he arrested, put in concentration camps, or killed about fifty thousand. Thus, he turned all their still-free relatives and neighbors into enemies and lost what they might have contributed to the effectiveness of his regime.

Even more important was his campaign against the peasant villagers who grew Vietnam's food. Reacting to his policies toward them, they would shelter the insurgents and eventually themselves become insurgents. I have described the status and role of revered village councils. They were dangerous, Diem decided, because, if elections were free, a "large number of Viet Minh might win office." So in June 1956 he abolished elections and replaced native, mainly Confucian, village notables with alien, mostly Catholic, appointed officials. "Villagers found them 'strange, and not a little incomprehensible,'" wrote the editors of the *Pentagon Papers*. Also, "since the officials were the creatures of the province chiefs, corruption at the province level . . . was transmitted directly to the village . . . [Thus,] their alien presence in the midst of close-knit rural communities encouraged revival of the conspiratorial, underground politics to which the villages had become accustomed during the resistance against the French." Worse were the troops sent by the government to win back the 60 to 90 percent of the villages the French believed to be controlled by or favorable to the Viet Minh. American observers saw them as "poorly led, ill-trained, and heavy-handed, the troops behaved towards the people very much as the Viet Minh had led the farmers to expect . . . symbol[s] of insecurity and repression." By 1962, about seven in each eight regional senior officials were soldiers, and the villagers were more afraid of them than of the insurgents.

Subversion of the traditional village way of life did not stop with the appointment of officials. Diem had come to see the villagers themselves as the enemy. They were too many to kill or incarcerate in concentration camps, but, drawing on the British experience in Malaya and Kenya and the American experience in the Philippines, he ordered that they be relocated to constitute "a 'living wall' between the lowland centers of population and the jungle and mountain redoubts of the dissidents. Between April 1957 and late 1961, the GVN [the government of South Vietnam] reported that over 200,000 persons—refugees and landless families from coastal Annam [the central area of Vietnam]—were resettled in 147 centers carved from 220,000 acres of wilderness."

This program was not merely "relocation"; it had an impact in Vietnamese culture far deeper than mere movement. As Frances FitzGerald writes in her sensitive study of the war, *Fire in the Lake*:

> In the old ideographic language of Vietnam, the word *xa,* which westerners translate as "village" or "village community," had as its roots the Chinese characters signifying "land," "people," and "sacred." These three ideas were joined inseparably, for the Vietnamese religion rested at every point on the particular social and economic system of the village. Confucian philosophy taught that the sacred bond of the society lay with the mandarin-genie, the representative of the emperor. But the villagers knew that it lay with the spirits of the particular earth of their village. They believed that if a man moved off his land and out of the gates of the village, he left his soul behind him, buried in the earth with the bones of his ancestors. The belief was no mere superstition, but a reflection of the fact that the land formed a complete picture of the village: all of a man's social and economic relationships appeared there in visual terms, as if inscribed on a map. If a man left his land, he left his own "face," the social position on which his "personality" depended.

Thus, in uprooting and dispossessing villagers not only from their physical possessions but also from their spiritual roots, the

Diem government struck not only at their physical well-being but also at their spiritual existence. While they could not fight back effectively, the villagers were horror-struck. It was their shock and anger that prepared the way for the Viet Minh.

In less than a year after Diem's relocation program began, the Viet Minh were able to mount an organized, widespread, and popularly supported rebellion. In this rebellion, the Communist government of the North does not seem to have played a significant role. In fact, it attempted to dissuade the "stay behinds" in the South from precipitating action against the Diem regime. Hanoi was not seeking compromise; it simply felt that the ground had not been prepared for guerrilla war. It ordered the Southern Liberation Armed Forces (Giai Phong Quan) to spend their time and energies building their organization. American intelligence reported that "most of those who took up arms were South Vietnamese, and the causes for which they fought were by no means contrived in North Vietnam." Their move fit within a pattern of insurgency.

In my study of guerrilla warfare, as I point out in the introduction, I found that the "pattern" of insurgency involves three stages. The first is political. The insurgents must establish their right to speak for the nation. They usually do this by opposing the intruder. In Vietnam, Ho Chi Minh did not achieve popular recognition as leader of the Communist Party (the Lao Dong) but as a nationalist, an opponent of the foreigners who dominated his country. In effect, he carried the Lao Dong on his coattails rather than the reverse. So, primarily through his well-publicized nationalist role and eventually through the superb organization he built village by village, at least Ho and through him the Lao Dong had won the political component of the war long before the Americans arrived on the scene.

The second stage of insurgency involves destroying the administration of the enemy and substituting for it the administration of the insurgents. This is what the southern branch of the Viet Minh, the Giai Phong Quan, did: it swept away the Diem-appointed administration by assassinating the appointed village officials. The well-informed French observer Bernard Fall estimated that Giai Phong

Quan militants killed about seven hundred of these detested officials during 1957–1958. The numbers rose dramatically thereafter: Fall estimated twenty-five hundred from 1959 to 1960 and four thousand from 1960 to 1961. But it was not just the officials who were liquidated. As George Carver of the CIA wrote in 1966 in *Foreign Affairs*, "The terror was directed not only against officials but against all whose operations were essential to the functioning of organized political society, school teachers, health workers, agricultural officials, etc." One captured document orders cadres to "carry out assassination missions right at the center to immobilize the enemy. Prime targets should be security forces and civil action district officials, hooligans and thugs. Besides propaganda under armed protection must be carried out on a regular basis with a view to establishing bases."

Then, having either incapacitated the rival government or killed its representatives, the Viet Minh set out to create their replacement as administrators of the territory it held. What it did was classic. We have seen in the American Revolution, towns and cities all over the colonies created Committees of Safety to deliver the mail, print money, and administer courts of law; in Spain, guerrillas assessed and collected taxes, operated customs posts, and formed city, regional, and national *juntas* that effectively replaced the monarchy; in Yugoslavia, the Partisans printed newspapers, manufactured arms, and even ran a railway; in Greece, the EAM organized village councils that collected taxes, ran schools, and adjudicated local disputes. In practical, day-to-day affairs, all these groups became the government.

When I did my original study of guerrilla warfare in Vietnam in 1963, quantification was in favor. So I weighted the relative importance of phases of guerrilla war. I concluded that the political element was overwhelming, say 80 percent of the total. Elimination of the existing administration and creating a new administration was perhaps 15 percent. That meant that the military component—which was all that remained when America entered the war and on which American thought and action concentrated—amounted to only the residual 5 percent. That is, America seized the short end of the lever.

Ho was not keen to move from the second (administrative) to the third (military) stage of the insurgency. Indeed, initially, he "did not expect to have to resort to force." But the Southerners got out in front in 1958, contrary to the instructions from the party. So worried was Ho by their precipitous action that he sent a member (later first secretary) of the Lao Dong politburo, Le Duan, on a fact-finding mission to the South. When he returned to report that the movement was serious, Ho decided to support it. "That fall [1958] when the monsoon rains ended and the trails through Laos started to dry," Neil Sheehan wrote, "the first few hundred infiltrators marched south. They were the first of the thousands to follow in the coming years from among the Southern Viet Minh who had been withdrawn to the North after Geneva . . . Soon the fighting forces quadrupled to 10,000." At the time, neither Diem nor the Americans realized that a major shift had occurred: "Not until 1960, however, did the U.S. perceive that Diem was in serious danger of being overthrown and devise a Counterinsurgency Plan . . ." according to the editors of the *Pentagon Papers* after a review of all the intelligence sources. By the time President Kennedy decided to intervene, in November 1961, the insurgents in the South had reached more than sixteen thousand and had captured at least eleven thousand weapons. To magnify their power and increase their appeal, they formed an alliance with non-Communists. Thus, they kept the pretense of a separate Southern movement, but as the Communist Party secretary-general later said, "Officially, we were separate, but in fact we were the same thing all the time; there was a single party; a single government; a single capital; a single country."

In fact, the unity of the movement was its most striking feature. The Viet Minh was dominated by the Communist Party, which itself was dominated by a small elite. As the editors of the *Pentagon Papers* write, "This group of leaders were unique in the communist world for their homogeneity and their harmony—there has been little evidence of the kind of turbulence which has splintered the leadership of most communist parties. While experts have detected disputes within the Lao Dong hierarchy—1957 appears to be a critical year in that regard—the facts are that there has been no blood-purge of the Lao Dong leadership, and except for changes occasioned by apparently natural deaths, the leadership in 1960 was virtually identical to what it had been in 1954 or 1946. This remarkably dedicated and purposeful group of men apparently agreed among themselves as to what the national interests of the DRV [the Democratic Republic of Vietnam] required, what goals should be set for the nation, and what strategy they should pursue in attaining them." This explains why the American government could not apply "the lesson of Greece" to Vietnam: whereas the Greek Communist Party split and so virtually committed suicide over its relationship to the Soviet Union, the Lao Dong remained united. Nothing America did could divide it.

By 1961, the American government believed that the Viet Cong (as the Giai Phong Quan or Liberation Army in South Vietnam had been renamed in the press) numbered about 16,500, of whom about half, who were organized along standard army lines into companies and battalions, were "main force." The other half were regional and local forces, who were daytime farmers, fought as guerrillas whenever they had a "military advantage of at least four to one," and constituted a manpower pool on which the main units could draw. Most of the leaders of these two forces were originally from the South and had gone north during the exchange of population after the Geneva Accords. "Under their leadership, large areas of the South became mini-states where taxes were levied, training facilities built and fortifications dug." The military buildup was important and rapidly accelerated, but even more important was the Viet Minh political campaign, particularly the linking of the increasingly feared and hated government of the South with the United States.

As the editors of the *Pentagon Papers* write, emphasizing that their opinion was based on all sources of intelligence available to the U.S. government, support for the Saigon government among South Vietnam's peasants—90 percent of the population—"was weak and waning . . . peasant resentment against Diem was extensive and well founded. Moreover, it is clear that dislike of the Diem government was coupled with resentment toward Americans. For many Vietnamese peasants, the War of Resistance against French–Bao Dai rule never ended; France was merely replaced by the U.S. and Bao Dai's mantle was transferred to Ngo Dinh Diem. The Viet Cong's op-probrious catchword *'My-Diem'* (American-Diem) thus recaptured the nationalist mystique of the First Indochina War, and combined the natural xenophobia of the rural Vietnamese with their mounting dislike of Diem." To the contrary of the later slogan, Vietnamization, the war in 1961 had been Americanized.

No one in the American government knew what to do about this different kind of war. The military thought of it essentially in terms of what they had experienced in Germany and Korea: main battle groups fighting it out on battlefields. But battlefields did not exist in Vietnam, and the enemy forces were scattered small groups that disappeared when confronted. Vietnam in effect was Spain redux, with Americans beset as was Napoleon's splendid army by elusive guerrillas. Worse, until later in the 1960s, Americans could only "advise" and had to rely upon the South Vietnamese army, whose officers were usually intent on avoiding combat. American attempts to "stiffen" Vietnamese units with Special Forces troops did not work. So, by the end of the decade, as Neil Sheehan wrote, "the American commander, General William Westmoreland, was intent on shunting the Saigon forces out of the way so that he could win the war with the U.S. Army." As the Americans built up their combat forces, so did the Viet Minh, which increasingly brought regular formations south from North Vietnam. What had been squads became companies, and companies grew into regiments.

The CIA thought it had better answers than the army leaders. The answers the CIA brought forward were grouped together under the rubric that became the buzzword of the 1960s, counterinsurgency. For models, it drew upon experience in the Philippines, Greece, and Malaya. From Malaya, under the guidance of Robert Thompson, a British officer who had played a role there, it hit upon the idea of isolating the population so that it could not support the guerrillas. That was the essence of the concept of strategic hamlets.

The strategic hamlet program (known in Vietnamese as *ap-chien-luoc*), under a different name, had already been tried between 1957 and 1961 when some two hundred thousand people were moved to

147 centers in remote locations. This program had infuriated the peasants who were expelled from their villages and facilitated the advent of the Viet Minh, but, instead of dropping it, the Saigon government and its American advisers planned to enlarge it to encompass eleven thousand of the roughly sixteen thousand hamlets in South Vietnam. By June 1963, an estimated 67 percent of the rural population was living in some sixty-eight hundred barbed-wire-encircled strategic hamlets. As the editors of the *Pentagon Papers* write, "The strategic hamlet program was, in short, an attempt to translate the newly articulated theory of counterinsurgency into operational reality. The objective was political though the means to its realization were a mixture of military, social, psychological, economic and political measures . . . The long history of these efforts were marked by consistency in results as well as in techniques: all failed dismally." As Neil Sheehan observed in trips through the countryside, by the end of 1963, "Most of the thousands of strategic hamlets that [U.S. Commander General Paul] Harkin listed on his [briefing] charts ceased to exist." As quickly as they could, the inhabitants went home.

The counterinsurgency plan on which the strategic hamlet program was based called for the hamlets and the remaining villages to be protected by the South Vietnamese Self-Defense Corps (SDC) and the Civil Guard (CG). But these units were so "poorly trained and equipped, [and] miserably led [that] they could scarcely defend themselves, much less secure the farmers. Indeed, they proved to be an asset to the insurgents in two ways: they served as a source of weapons; and their brutality, petty thievery, and disorderliness induced innumerable villagers to join in open revolt against the GVN [government of South Vietnam]." As the program broke down, Sheehan found in his travels around the South that the "guerrillas hardly bothered anymore with the small posts and tiny watchtowers. Their garrisons fled, and those that stayed and survived did so because the Viet Cong left them in place to serve as quartermasters for fresh ammunition. A standard price for a month's survival was 10,000 rounds. The demoralized militiamen would turn it over and requisition 10,000 more to survive next month by telling the district chief that they had been attacked and fired it off themselves." When the American officials attempted to upgrade these forces so that they could better defend themselves, their effort simply served to supply the attacking guerrillas with modern automatic weapons to replace the crude shotguns they had manufactured themselves and the older rifles they had captured from the French. So plentiful were American weapons—in the hundreds of thousands—that the insurgents stopped using Soviet equipment.

Since the main battle formations of the U.S. Army could rarely pin down and destroy the increasingly large and sophisticated insurgent forces and the strategic hamlet program had failed, the CIA convinced the Johnson Administration in 1967 to engage in a program of targeted assassination. The "Phoenix Program" (also known as Civil Operations and Revolutionary Development Support, CORDS) would result in the imprisonment, torture, or death of tens of thousands of South Vietnamese. Inevitably, like everything else in South Vietnam, it became a gold mine for corrupt officials: even known terrorists could buy their way out of the clutches of the operatives while many innocent people were forced to ransom themselves. Its effect on the course of the war was, to say the least, ambiguous.

Meanwhile, the Viet Minh pressed steadily into the South. As they had done before Dien Bien Phu, they threw their massive manpower reserves into the construction of infrastructure. The Ho Chi Minh Trail was built and maintained by upward of half a million people. Repeatedly, indeed almost daily destroyed (as I saw on the aerial surveillance images), it was also daily rebuilt and even improved so that by the late 1960s, its multiple, parallel trails had become thousands of miles of roads able to support trucks even in rainy weather. Despite constant bombardment, there was an "undiminished flow of men and supplies." This flow was augmented, particularly for fuel, by a virtual fleet of tankers and barges running ashore along the fifteen-hundred-mile coastline. Despite the highly sophisticated equipment available to the Americans, they were not able to interdict either supply route.

Finally, on January 31, 1968, Ho's military commander, Vo Nguyen Giap, was ready to strike. First, he duped the American command into believing that his forces were about to attack the fortified base of Khe San. To the American command, it seemed a replay of Dien Bien Phu. They were delighted. The attack would be fatal to the Viet Minh since, unlike the French, the Americans had mobility, firepower, and vast human resources. Of course, Giap knew this so he had no intention of attacking Khe San. With remarkable secrecy, he brought together some seventy thousand regular troops who were led to their targets by Southern guerrillas. The Tet offensive was a stunning operation.

Militarily the Tet offensive was nearly a disaster for the Viet Minh, as many thousands of their troops were killed, but politically it was the masterstroke that won the war. After Tet, the American public had no stomach for Vietnam. It even brought about a "regime change" in America as Lyndon Johnson withdrew his candidacy for the presidency. Thereafter, the war wound down by fits and starts, with more emphasis on punishing aerial attack, which aimed ostensibly to create a "negotiating climate." That policy had a hard edge for America: during the withdrawal period an additional twenty-one thousand American soldiers were killed. Thus, the American phase of the war that began in 1945 ended in the Communist victory on January 27, 1973. It was the longest, most brutal, and most destructive guerrilla war in modern history.

CHAPTER 11

THE AFGHAN RESISTANCE
TO THE BRITISH
AND THE RUSSIANS

REMOTE, RUGGED, AND BARREN, SOMEWHAT RESEMBLING IN SIZE AND terrain a combination of Colorado and New Mexico, Afghanistan was one of the few pieces of Asia and Africa not incorporated into the European empires in the nineteenth century. From the earliest recorded history, it has been a route rather than a destination. Chinese and Indian Buddhist pilgrims struggled over its high mountains passes; Alexander's Macedonians fought their ways through its mountains and deserts down to India; Central Asian Turks and Mongols sang of the delights of its rare gardens and limpid air but rarely tarried; and the Russians and British used its craggy heights and boulder-strewn valleys only to play their "Great Game" of espionage and cold war against each other. The British tried three times to add it to their Indian empire before giving up. Why, in 1979, long after the British had withdrawn from South Asia, the Russians neglected history and sought to conquer Afghanistan still baffles both Russian and Western students of strategy. As one former Russian intelligence (KBG) officer shook his head in perplexity, "why Afghanistan? We have enough mountains in the Soviet Union already." But there was an ironic logic in the Russian policy: it was the Russian version of the domino theory that so worried American strategists, but in reverse. Instead of worrying, as did John Foster Dulles, that the fall of Vietnam would spread Communism to surrounding countries, Leonid Brezhnev feared that a Communist failure in Afghanistan would impact on the Tajiks, Uzbeks, and other Turkoman peoples of Soviet Central Asia, infecting them with anti-Communist aspirations.

For the Afghans, of course, what the British and Russians thought—and sought—was irrelevant. They just wanted the foreigners to leave them alone. They always had stoutly opposed the entry of foreigners; the struggle against them was the substance of their sagas and myths. They hated the foreigners even when, as was true of many of them, they counted the foreigners among their forebearers. The people of the southeastern area known as Nuristan swear that they are descended from Alexander's legions, and the Hazaras of the high Hindu Kush mountains count Genghis Khan's hordes as their ancestors. More recently, many of the grandfathers of Afghanistan's Tajiks and Uzbeks were refugees who had fled the Communists during the bloody civil war that followed the Russian Revolution. More than almost any other society, Afghans live their history. So, before looking at the guerrilla war of the 1980s, I begin with where they begin, the Great Game of the nineteenth century.

As the British thrust across the Indian subcontinent, collecting one after another the petty states into which the Mughal Empire had shattered, they reached what is now Pakistan by 1820. To the west was Sind, where they cowed a motley collection of local rulers into a treaty in that year. The treaty aimed to exclude European traders and—astonishingly—American settlers. European traders conceivably might come to compete for the market, and the British, in the middle of their great commercial expansion, were sure they would, but American settlers were a figment of their inflamed imagination. With Sind in their hands, the British turned northeast on the Indus river toward the Punjab, where in the last year of the eighteenth century, a remarkable Sikh leader named Ranjit Singh had begun to create an empire to rival the British. His was a forlorn ambition. Nothing could stop the British, and upon his death in 1839, they incorporated his territory. Control of the Indus valley brought them into contact with the Pathan peoples in what they later called the Northwest Frontier and Afghanistan.

Always moving forward, the British found no secure frontier. They were already beyond the Indus river and were still far from the mighty wall of the Hindu Kush mountain range. Beyond each hill was another valley, and there seemed no place to draw a line. That mattered to them because, within the living memory of many of their generals, other Europeans were trying to do what they had already done, conquer India. Napoleon got as far as Egypt before turning back. His attempt was more sound than fury, falling as it did more than a thousand miles too short, but the forward movement of the Russians lasted far longer and could be painted on the map.

The Russian advance began with Ivan IV, "the Terrible," who in Russia's first great military adventure in 1552 conquered a remnant of the vast Mongol Empire, the Khanate of Kazan. It was the conquest of Kazan that began the transformation of Russia from a petty city-state into a multinational empire. Because Kazan was an Islamic society with elaborate political, commercial, and intellectual structures and a fully formed legal and religious code, Ivan's churchmen set out on a thoroughgoing regime change, "to convert the pagans to the faith." And because the society was multiethnic, each of the newly defeated groups presented a key to further annexation. Next in line was the Khanate of Astrakhan, whose conquest in 1556 opened the Volga down to the Caspian Sea.

Ivan the Terrible already understood that the main obstacle to his imperial dream was England, so, as Winston Churchill was later to propose to Josef Stalin, he offered England a grand compromise: if Elizabeth, the Virgin Queen, would marry him, they could exchange wedding gifts that would meet the needs of both countries. Ivan would give the British a commercial monopoly in Central Asia if the

British would give him the arms he needed to solidify his conquest of it. The queen declined. Furious at this slight on his royal dignity, Ivan wrote, addressing her as *poshlaia dvitsa* (common wench); so with tongue in cheek we could date the beginning of Russo-British rivalry over Central Asia.

Elizabeth had sound dynastic reason for her rebuff of Ivan: Muscovy was still a minor state. It had overreached its power. Another of its Muslim neighbors, the Crimean Tatars, were able to regroup. They continued to fight the Muscovites and in 1571 even burned Moscow's suburbs, kidnapping and enslaving thousands of Russians. The Russians had lived in terror—and grudging admiration—of Turkic and Mongol peoples for generations, and in their first significant diplomatic contacts with the Muslim state of Kazan had come to respect its social and cultural achievements. Consequently, once a Central Asian, loosely collected under the term "Tatar," converted to Christianity, he was completely accepted into Moscow society; some Tatars eventually married into the Russian royal family, while those of noble background were often awarded the title *tsarovich*, or prince. The Russians never developed the sense of racial superiority the British evinced in India.

After Ivan died, Muscovy focused on domestic and European affairs. Then in 1721, Tsar Peter, "the Great," having made peace with Sweden, assembled an army and set off, as he said, toward the first stage on "the road to India." He planned to move down the Caspian to Herat on the border between Iran (or, as it was then known, Persia), then turn east to the ancient fortress city of Ghazni, turn up the Helmand river to Kabul, and cross the Khyber Pass south to Lahore. He did not make it. His troops were attacked and nearly annihilated by Turks from the Khanate of Khiva. Failure though it was, it was encapsulated into the long-lasting British nightmare of a Russian invasion of India.

Meanwhile, a young man whom Peter would probably have liked personally, a Ghilzai Afghan warrior named Mahmud, had made himself lord of the province of Qandahar. Instinctively he knew that the way to the hearts of the tribesmen was the capture of booty, so he led a *lashkar* (raiding party) of Afghans across the forbidding Dasht-i Lut desert to sack the prosperous caravan city of Kerman. At that time, Safavid Persia, like Mughal India, was dying from the top. As the shah withdrew into his harem, the government fell apart and the provinces rose in revolt. Peter watched this disarray with interest. When a group of Russian merchants were robbed by outlaws, he saw an excuse and was tempted to move forward. But Mahmud beat him to the prize, Isfahan—then a city about the size of contemporary London—and forced the ruling shah to surrender. Mahmud's capital, Qandahar, which was the seedbed of modern Afghanistan, had become the center of a great empire determined and (temporarily) able to hold back both Russia and Britain.

Thus each group—the Afghans, the Russians, and the British—acquired a mindset that would determine their relations until our own times. The Afghans, while not yet a clearly formalized nation, had become imbued with a determination to rule their own neighborhood. The Russians, by the time of Catherine the Great, had come to feel a "manifest destiny" for the East as surely as Americans would for their West. Catherine resumed the march south and east. By 1792 the Russians had overwhelmed the Crimean Tatars and a few years later moved steadily, petty Turkish state by state, down the shores of the Black, Caspian, and Aral seas toward Persia and Afghanistan. Their route of march would lead them, the British believed, toward the goal Peter had proclaimed—India. And, obsessed as they were with fears of hordes of Cossacks galloping down from Russia into their new Indian empire, the British believed they would have to defend India from the Russians even if that meant having to conquer the Afghans.

Conquering the Afghans was attractive because their territory contained the only defensive line between the Indus river and Russian-dominated Central Asia. The mighty Hindu Kush mountain range, in the middle of Afghanistan, would then become a no-man's-land. In fact, it did. It became the playing field of the "Great Game" for both British and Russian officers and spies. One of the Russians, Captain Yan Vitkevich, who visited Kabul to offer its ruler a pact with Russia, electrified the British authorities in India. Their first reaction was to send an ultimatum to the Afghan king Dost Mohammed demanding that he "desist from all correspondence . . . with agents of other powers." But they could not be sure that Dost Mohammed, whom they thought of as "wily and deceitful," would obey their command; so they decided to push into Afghanistan to put it under a puppet regime. That key move in what they then called the forward policy, they thought, would definitely and permanently checkmate further Russian advance. That was the cause of the First Afghan War in 1838–1839. On paper, it was a shrewd strategy, but it neglected one element, the Afghan people.

As in the many contemporary wars in Asia and Africa, the disciplined European troops and their native auxiliaries armed with cannon overwhelmed the Afghan forces, who had neither discipline nor cannon. In August 1839 the British captured Kabul and installed their puppet ruler and settled down to organize a new regime. The regime was hated. As the late Louis Dupree wrote, "The short, unhappy reign of [the British-installed ruler] Shah Shuja began, propped up by British bayonets, supported by British gold, sustained by British and Indian blood. By most contemporary accounts, Shah Shuja, never popular with his people, encouraged further enmity as the glaring presence of the *farangi* [British] bayonets increased hatred and distrust." From military triumph, the British sank into an insurgency they could not quell. As the historian Stanley Wolpert has written, "Guerrilla warfare, assassination, local uprisings, and looting became daily occurrences wherever British Indian troops were found in the land of the Afghans, in the bazaars of Kabul and Kandahar [Qandahar], along the road to the Khyber, in the palace itself." The British were baffled. Their commander, General Sir William Macnaghten, then wrote, "I have been striving in vain to sow *nifak* [dissension] among the rebels and it is perfectly wonderful how they hang together." What held them together and inspired them, the British decided, was a combination of Islam and hatred of foreigners. That insight was both an assessment and a prophecy.

The British faced the classic dilemma of occupying powers:

while it had been relatively easy to move in and overwhelm the army and government, it was increasingly difficult to move out. "Flexible response," the cliché coined in our own times, is easier done than flexible "unresponse." For the British in Kabul the problem quickly became acute—if they withdrew, their protégé would be chased away and their enemies would return, perhaps with Russian help, so the situation would be worse than before. Their defeat would encourage their enemies and terrorism might spread to India itself. But staying the course was expensive, unpopular, and insecure. The British solution was the usual mistake of governments, compromise, and produced the worst of both policies. They withdrew a large part of their contingent, mainly the fighting force, thus cutting down on expenses, while showing their decision to stay by importing the wives and families of their officers and the families and camp followers of their Indian troops. Almost overnight a virtual new city, a fortified enclave in Kabul—an early version of the "Green Zone" the Americans later created in Baghdad—sprang up haphazardly to house them all.

As attacks mounted, the British dallied, unable to stay but afraid to leave; then at the worst possible time, in the winter of 1841–1842, they abandoned Kabul and began to retreat down the Khyber Pass. Of the 16,500 people (including 4,500 troops) who set out, only one survivor reached British-held Jalalabad. The first British attempt to "pacify" Afghanistan had ended in the worst disaster the British army suffered in the nineteenth century, and the Afghans had won the most impressive guerrilla war of the century.

The war left legacies that have dominated Afghan history to the present day. The most important to outsiders was the recognition of the pivotal role of Afghanistan in Asian affairs. Never again would the British and the Russians be able to leave it alone. Viewing the British defeat—the Afghan version of Braddock's disaster during the French and Indian wars and the French defeat at Dien Bien Phu—the Indians were encouraged to believe they too might be able to drive out the foreigners. Fifteen years later, in the great Indian "Mutiny," they would try. As a contemporary British officer, unconsciously echoing Benjamin Franklin, laconically commented, Afghans and Indians "had perceived that we were not invincible." Afghans had shown that imperialism could be defeated or made too costly to sustain, but their victory had been Pyrrhic: Afghan society, already fractured by culture, race, and religion, had been further split into mutually hostile communities so that subsequent attempts at reform and nation building would fail. As Afghans, Britons, Russians, and Americans would learn in the years to come, the guerrilla genie, once it had escaped from the bottle in 1840, has refused tamely to return.

Meanwhile, Britain retaliated for its humiliation as best it could, mainly with search-and-destroy raids, but for a generation it was occupied with absorbing Sind, Punjab, and the numerous Indian petty states just as the Russians were doing in Central Asia with Bukhara, Khiva, and Kokand. The two great powers were converging on Afghanistan.

Why they were doing so was puzzling. But the tsar's foreign minister, Prince A. M. Gorchakov, in December 1864 offered an explanation that contemporary British and French statesmen would have found persuasive:

> The position of Russia in Central Asia is that of all civilized states which are brought into contact with half savage, nomad populations possessing no fixed social organization. In such cases it always happens that the more civilized State is forced, in the interests of security of its frontiers and its commercial relations, to exercise a certain ascendancy over those whom their turbulent and unsettled character makes undesirable neighbors. [To do so, it must advance] deeper and deeper into barbarous countries . . . Such has been the fate of . . . the United States in America, France in Algeria, Holland in her colonies, England in India, all have been irresistibly forced, less by ambition than by imperious necessity, into this onward movement where the greatest difficulty is to know where to stop.

Not knowing where to stop got Britain into the second Afghan war. The "lesson" of Afghanistan had been forgotten by the next generation of Englishmen, but the Afghans had not forgotten. They reacted with anger and fear to the buildup of a large British diplomatic and military staff in Kabul. Finally, in 1879, they assassinated a number of British officers. In their turn, the British grew fearful that the Russians might use the British humiliation to intervene. That was the general cause of the British decision to send a second army into Afghanistan. Since the only result was that the Afghans agreed that Britain would guide its virtually nonexistent foreign policy, the war produced no useful result, but so brutal were its exploits as to cause the fall of the English government in London and the replacement of its senior officials in Delhi.

The Russians were elated. Some even began to think that the sun of the British Empire was setting. Their hopes seemed confirmed when another Muslim force, the followers of the Mahdi in the faraway African Sudan, routed a British army and collapsed the incipient British empire in central Africa. With the encouragement these events gave them, the Russians again moved forward, this time assisted by the building of a railroad that seemed to the British like an arrow pointed south toward India. At the otherwise unimportant oasis of Panjdeh, a dependency of Afghanistan north of Herat, the Russians attacked and virtually annihilated a ragtag army of Afghans in March 1885. The British didn't care, of course, about the Afghan losses or about Panjdeh, which presumably few of them could find on a map, but they saw Panjdeh as another stepping stone on the route to India. Alarmed, Parliament voted a war chest. It seemed that the Crimean War was about to be refought.

Both sides recoiled from this near collision to deal with more urgent issues in Europe. Britain agreed that the Russians keep much of the area around Panjdeh, hoping it would be the final Russian demand. The two powers moved to agree on frontiers—a Russo-Afghan frontier along the Amu Darya river and an Indo-Afghan frontier in the foothills of the Hindu Kush mountains. A few years later one of the most desolate areas on the planet, the high mountains and frozen valleys in the far northeast of Afghanistan, was made into a buffer zone known as the Wakkan Corridor. On the map of what the British statesman Lord Curzon had called the "chessboard upon which is being played out a game for the domination of the world," no further moves were to be allowed.

While the rest of the world plunged into the First World War, the Afghans enjoyed a rare period of tranquillity. A progressive new ruler proposed a program of reforms designed to make Afghanistan less of a chessboard for Europeans and itself a more capable manager of its own affairs. Alarmed rather than reassured by the fall of their old enemy, the tsar, the British reacted sternly when King Amanullah tried to reassert Afghan power over the Pathan tribes on what the British called the Northwest Frontier. That was the cause of the third Anglo-Afghan war in the spring of 1919. Again, the British won all the battles but lost the war. In the treaty ending the fight, they gave up control over Afghan affairs and even allowed the Afghans to establish relations with the new Communist government of Russia.

The Great Game was in remission for a generation. It seemed to have ended definitely when the British withdrew from India in 1947. But what actually happened was that Britain's role was taken up, haltingly, almost inadvertently, by the Americans as a sort of adjunct to what really interested the Eisenhower Administration, the countries that Secretary of State John Foster Dulles had drawn together (including the newly established Pakistan) into what was popularly known as the Baghdad Pact.

It was at the beginning of the Kennedy Administration that I became involved with Afghan affairs. There was very little, I found, in the government archives and nothing in the press worth reading about Afghanistan. So I was delighted when the then undersecretary of state, Governor Chester Bowles, asked me to join him for an inspection tour. What we found was that the tiny American embassy was considered a hardship post that ambitious officials sought to avoid. The AID mission director had never been outside the capital, and his team took a relaxed view of the ill-conceived projects that had been mounted in the previous administration. I decided to make an analysis of Afghan-American relations. At the invitation of the then Afghan government, I made a two-thousand-kilometer inspection tour by jeep that, among other things, introduced me to the provincial governors and tribal leaders. What became clear from the visit was that Afghanistan was a remarkably varied society on which the guidance of the central administration rested lightly. The people were among the poorest in the world but had a sense of independence and personal worth that virtually overcame their poverty. The Afghan code of honor, particularly but by no means uniquely practiced among the Pathan tribes where it was known as the Pukhtunwali, illuminated and guided their lives.

The Pukhtunwali rested on the concept that every village, clan, or tribe was a separate entity, virtually a miniature nation-state. Each had the collective obligation to defend its citizens, their property, and their honor. Thus, it absolutely commanded the taking of revenge (*badal*) for wrongs or insults to any of its members by outsiders, as the British had learned and the Russians soon would. As among the pre-Islamic Arabs, in the absence of overarching civic institutions and organizations, the certainty that revenge would be taken was the final, indeed the only, safeguard for the individual. That was the theory, but the practice was unending feuding. Consequently, every Afghan was armed and always ready to fight.

The imperative of revenge was softened by the parallel imperative of hospitality (*melmastia*). Afghan refugees or travelers could demand, and would receive, both hospitality and protection even from enemies. In villages where the inhabitants teetered on the brink of starvation, my team and I were greeted with ruinous generosity. To have attempted to pay for or to have refused what was offered would have been a mortal insult. As guests we were under the protection of our hosts, who had the absolute obligation to defend us or die trying. (Americans would later be baffled by the silent refusal of such poor people to turn over Osama bin Ladin for the, to them, astronomical sum of $25 million.)

Also evidently governing the lives of Afghans was their religion, Islam. Islam did not become a part of the life of the Afghans until the tenth century, more than three hundred years after it was almost universally adopted by Arabs, Persians, and Berbers. But, in every moment of the day, as I observed, it regulated life and, like Christianity, Hinduism, and Judaism, incorporated local custom and belief. More, it permeated the whole society. Any person who felt the urge to talk about, teach, or lead his fellows became in his own eyes and those of his fellows what Christians would think of as a priest or pastor, and Jews, a rabbi. The small and diverse Afghan society included at least a quarter of a million mullahs. Being a mullah, however, did not make a man any less a herdsman, farmer, merchant, or warrior. (The Russians would later learn this to their great cost.) Even apart from the mullahs, the Afghans evinced a deeper commitment to Islam in their daily lives than I had ever seen in my years of study and residence in the Arab countries. Throughout history in times of crisis or warfare, this commitment to religion is manifested in the requirement to perform jihad—religious struggle including, if necessary, holy war. Not just mullahs but every man was expected to serve as a *mujahid* or voluntary defender of the faith even if, indeed especially if, doing so was likely to force him to die as a "martyr" (a *shahid*) for having "borne witness" (*shahada*) to his faith. Afghan guerrillas, later fighting against the "Godless Russians," would call themselves mujahideen.

Two other partly contradictory forces were evident everywhere I went: the dominance of recognized headmen, usually known as *maliks,* and the insistence on a form of primitive democracy, the popular assembly, the *jirga.* The *jirga* is the occasion where the opinions of the community are sorted out, and the *malik* becomes the manifestation of community consensus. The writ of the *jirga* is local, and the *malik* usually has little influence and no power outside his own group. This autonomy set the pattern that was later manifested in the guerrilla war against the Russians. There was no effective way that villages could be grouped together and so coerced by the Russians, but also no way that they could become a single political or guerrilla organization to fight the Russians. Just as there were about fifteen thousand villages, each with its own *jirga,* so would there be in the 1980s thousands of groups of men who functioned at least part-time as guerrillas.

The one institution that gave partial unity to Afghanistan was the monarchy, but the figure who embodied it was only temporary. The king who won Afghan independence in 1919, Amanullah, was overthrown by a tribal rebellion ten years later. During the period of chaos that followed, power was briefly seized by an illiterate Tajik bandit chief who was a prototype of the warlords who virtually destroyed Afghanistan sixty years later in the 1990s. After a period of savage repression and organized mass looting, a member of the old ruling order, Nadir Khan, rallied the Pathan tribes. The price Nadir had to pay was that the forces on which he had to rely, "several thousand tribesmen, undisciplined and hungry for loot," had to be allowed to sack Kabul. Then, in a ceremony like the ones in which Roman legions "elected" their emperors, the tribesmen brandished their weapons and shouted Nadir Khan onto the throne. And, like many of the Roman emperors, Nadir himself would be assassinated after a brief reign.

Kings followed in rapid succession, but the monarchy as an institution survived until 1973, when King Mohammed Zahir was overthrown by his brother-in-law, Sadar Muhammad Daoud, who proclaimed what he euphemistically called a republic. Like many Afghan rulers, Daoud alternated reform programs with tyranny and managed in the course of his rule to alienate both the young liberal reformers who had grown in the permissive atmosphere of Zahir's rule and the conservative mullahs who represented traditional Afghanistan. In retrospect, however, the key domestic aspect of Daoud's period of control, which lasted only five years, was that it removed the unifying force of the monarchy while the key foreign aspect was that it brought back, in a new form, the Great Game.

Daoud, like his cousin, had invited Soviet participation in arming and training his army. There were then about three thousand Russian "advisers" in Afghanistan. They were closely watched by the Afghans and were careful not to appear to take part in Afghan politics. But obviously they had some impact on the attitudes of their "students." Some of the latter were in league with the leaders of the leftist political parties. Daoud grew suspicious of them and decided to arrest them. He bungled the move, and one managed to organize supporters in the army who besieged and then murdered him on April 27, 1978. They then proclaimed the "Democratic Republic of Afghanistan." Perhaps surprisingly, it was not a popular move because outside the army the Shravi, as the Afghans called the Russian successors to the British, were already universally unpopular. Needing them to survive, however, the new government invited the Soviets to double their number. Within a few months, Daoud's murderers fell out with one another, and the survivor, Hafizullah Amin, fearing for his life, invited the Russians to send in regular troops. The Russians later claimed that they had no choice but to comply, since without their help the pro-Russian government would have collapsed. Had it done so, they feared the ethnic conflict in Afghanistan might have spilled over into Soviet Central Asia. In the attempt to prevent this, Soviet troops began arriving on December 27, 1979.

So, within a year after the fall of Daoud, the U.S. government grew alarmed, as the British had been, by the Russian move into Afghanistan and what it feared might become a Russian move through Afghanistan toward India and Pakistan. A Soviet "forward" policy was alarming to the Americans, even though neither India nor Pakistan was integral to Western defense, and India, in particular, was often opposed to American policies. Perhaps more attractive to Washington was that, as a part of its cold war strategy, it was seeking ways to weaken the Soviet Union. As it did in other areas, it hit upon the idea of stimulating a proxy war against the Russians; with Pakistani and Saudi Arabian help, it began to identify, seek out, and supply with covert military equipment the Afghans who opposed the Russians. Such people were easy to find. Every village felt besieged by the Russians, who would ultimately number about 125,000. When some villagers reacted by stealing goods or shooting at soldiers, the Communist government or the Russians arrested or shot protesters; in panic, Afghans began to flee. This first major flight would become a torrent, of about half a million people. But, of course, most stayed; then as little groups of villagers got angry or hungry, they became guerrillas and began to kill Russians and their Afghan surrogates.

What might be called the fourth Afghan war was on. It would last a decade.

Most of the native opponents of the Russians were Muslim fundamentalists. Some were Sunni Muslims and others were Shiis. While all fought in the name of Islam, they fought separately by village or tribe. They were joined by foreign volunteers from Central Asia, the Middle East, Africa, and other Muslim areas. Of these volunteers, the best known is Osama bin Ladin, who came from Saudi Arabia. While the fundamentalists took American help and welcomed foreign aid workers and particularly doctors—and occasional journalists in the hope of creating favorable comment in the Western press—they made no secret of the fact that they hated with almost equal fervor both the Russians and the Americans. The Americans, in Afghan eyes, were just a new generation of Englishmen, intent on dominating their country and destroying their way of life.

The guerrillas were primarily villagers. The Russians, in the time-honored way of occupying powers, referred to them as "bandits" (*basmachis*). There never was a unified resistance organization. The first relatively large-scale guerrilla action occurred in the southeastern province of Nuristan. Attacks on the government and on the Russians then spread spontaneously village by village as *jirgas* met and as their *maliks* led their men in raids. Consequently, the groups began small—usually fewer than thirty men—and with a few exceptions stayed small. Some attempts were made at coordination, particularly through *chabnameh* (what have been termed "night letters"), which were passed from hand to hand within the mixed community of Kabul, denouncing the government and the Russians, but it does not appear that any serious attempt was made to create a "national" movement from the separate bands. Because of the history of the country, its poor internal communication, and its ethnic diversity, each group treated both the Russians and other opponents of the regime as rivals and often raided or fought them indiscriminately. As two of the most perceptive commentators, Richard Barnet and Eqbal Ahmad, wrote, the mujahideen were "too disunited to win the war, but they are too spread out to lose it."

It was only as increasing numbers of Afghans fled to Pakistan, numbering about two million by the spring of 1981, that they set up committees to act as middlemen with the Pakistanis, Saudis, and Americans. To convince their suppliers to support them, these "external mujahideen" attempted to coalesce into larger-scale resistance movements, and within the limits imposed by the Pakistani authorities, a dozen or so of them created virtual states within refugee camps and in such cities as Rawalpindi, each with its own armed followers, its arms merchants, and armorers. Thus, as in the Algerian war of the 1950s and 1960s, the Afghan resistance was divided into internal and external organizations. The actual fighting was done internally mainly by peasants led by their headmen, village by village, while the external organizations occupied themselves with collecting money, food, and weapons to support their followers. In effect, Mao's "sea" was being created in Pakistan while the "fish" remained in Afghanistan. The two groups were often cut off from each other. As Barnett R. Rubin then observed, "The commanders do not take orders from the exiled leaders. They derive their effectiveness and legitimacy not from party membership but from their ability to function as local leaders acting as intermediaries with outsiders." But communications, rudimentary though they were, were surprisingly good. Relying on runners to pass messages, they did not have radios whose coded messages, as in Vietnam, could be broken by the more sophisticated foreign power, but, as foreign visitors discovered, were quite effective even over long distances.

During most of the rebellion, each external "party" was restricted to a given ethnic group. Pakistan offered such coordination as existed among them; through Pakistan's intelligence service, American covert aid was channeled, while Saudi Arabia, also acting in coordination with the Americans, dealt with groups in touch with its own fundamentalist religious establishment. The leader of the Saudi group, Abdur-Rabb Sayyaf, who later would inspire Muslim guerrillas in the Philippines, became the first prime minister of the government-in-exile. To counterbalance this mainly Sunni Muslim grouping, the Revolutionary Government of Iran recognized and aided several Shia

Muslim émigré groups in Tehran, but they played almost no role in the internal resistance.

Unlike most of the insurgencies I have studied, the Afghan insurgency was motivated neither by nationalism nor by ideology. It defined itself in terms of its enemy. The enemy was not so much Communism as Russia, and not only Russia alone but all foreigners. The Afghans accepted outside help but did so reluctantly and without affection for the donors. Xenophobia must be considered to have been a major motivation. Insofar as it was refined into something like an ideology, it was defined by Islam. But it was not religion, per se, that seems to have most motivated people: it was the Afghan "way," the social code that was encapsulated in Islam, that Afghans felt was being attacked and that they determined to protect. As a Hezb-i Islami commander told the English visitor Peregrine Hodson, "It is true that Afghanistan is a poor country, but the most precious thing we have is our faith; without it we have nothing. We are fighting to protect our religion."

Dedication to an Islamic way of life, while obviously universal in Afghanistan, never proved strong enough to overcome ethnic differences or even geographical separation. The only insurgent organization that was able to overcome these barriers was what was originally and remained a largely Shia Tajik group that came to be called the Northern Alliance. It was formed by the charismatic guerrilla commander Ahmad Shah Massoud. He alone was able to draw adherents from neighboring communities. Like the leaders of other groups, he began with only about fifty followers. What set his movement apart, in addition to the fact that it grew and created alliances with other groups, was that it commanded a mini-state, the Panjshir valley northeast of Kabul. Even though they bombed, rocketed, and machine-gunned it from the air and repeatedly invaded with columns of tanks, the Russians were never able to conquer and hold it. Heavily outnumbered and outgunned, Massoud was driven in 1983 to work out a temporary cease-fire with the Russians. The cease-fire lasted a year, but unlike the deals with the enemy that crippled the Cetniks of Yugoslavia and undermined the EDES in Greece, Massoud does not seem to have been criticized for it. Rather he was praised for his cleverness in outsmarting the Russians and using the lull in fighting to rebuild his depleted forces, by then numbering about three thousand men, and to gather supplies to resume the fight.

Like many of the guerrilla leaders we have reviewed, Massoud wanted to convert his followers into a regular army. So, as he gathered strength, he levied taxes on the villagers and duty on gems mined in the nearby mountains; he also solicited contributions from Afghans in other areas. With these resources, he was able to pay his followers regular salaries so that they were able to fight full-time. He organized them into two kinds of formations: the local defenders were volunteers who were protecting their houses, families, and livestock. These *sabets,* as they were known, were essentially "minute men," at ready call. Being available was not difficult since the Soviet bombing campaign kept them restricted to caves and hideaways along the valley escarpment from which they could shoot down on the attacking Russians. Different from them were mobile units (known as *mutaharek*), with whom Massoud even attacked the Soviet airbase outside Kabul. Many of these fighters were not Massoud's Tajiks but had joined his forces because of his personal charisma and his reputation for success.

So disturbed by Massoud's growing importance were the Russians that when the truce expired in 1984, they attacked the Panjshir with overwhelming strength: Soviet air force jets carpet-bombed the entire valley as twenty thousand infantry troops aided by about five thousand Afghan government soldiers advanced behind columns of heavy tanks, and helicopters landed airborne contingents along the ridges. Anticipating the attack, Massoud had evacuated the thirty thousand inhabitants who had remained in the main valley and retreated into the side valleys that branched off it. As the perceptive French scholar Olivier Roy observed, Massoud shrewdly avoid combat and waited until the Russians were spread out and their lines of communication were overstretched. Then he counterattacked. The results were devastating to the Russians.

After the war, the Soviet general staff analyzed the reason for the defeat, the deaths of nearly fifteen thousand soldiers, and the wounding of another fifty thousand. They focused on the nature of guerrilla tactics that had been used against them. Since they had been made, painfully, into "experts," it is worth listening to their conclusions:

> Several combat principles lay at the heart of mujahadeen tactics. First, they avoided direct contact with the superior might of regular forces which could have wiped them out. Second, the mujahadeen practically never conducted positional warfare and when threatened with encirclement, would abandon their positions. Third, in all forms of combat the mujahadeen always strove to achieve surprise. Fourth, the mujahadeen employed terror and ideological conditioning on a peaceful populace as well as on local government representatives.
>
> The mujahadeen knew the terrain intimately, were natural scouts, and were capable of transmitting necessary information about secret Soviet unit and subunit movements over great distances using rudimentary communications gear and signaling devices.
>
> Among the guerrilla forces tactical strong suits were all types of night actions, the ability to rapidly and clandestinely move in the mountains, and the fielding of a very broad agent reconnaissance network.

That report might have been written about Tito's Partisans or Ho Chi Minh's Viet Minh.

During the fighting Medicines sans Frontièrs doctors observed more than six hundred Russian vehicles destroyed in the four provinces in which they worked. As Claude Malhuret, the organization's executive director, commented, "which, when extrapolated, comes to a total figure of some three to four thousand for the entire country." Even more impressive was that when ground-to-air missiles began to be made available, the mujahideen claim to have destroyed four hundred aircraft. The Russians virtually stopped flying, and lacking air cover that had tied down and discovered guerrilla forces, Russian ground forces were more vulnerable to ambush and tended to pull back to the cities. As a result, for most of the war, they occupied only about a fifth of the country.

Much as the Americans did in Vietnam in 1967 and 1968 and the Germans did in Yugoslavia, the Russians employed the most brutal forms of counterinsurgency: they aimed essentially to destroy the country and kill the inhabitants. Carpet bombing was used on towns and villages; dikes and irrigation works were blown up so that agricultural production fell by about half; forests were burned; roads and bridges were cut; and millions of small bombs were seeded into the countryside killing or wounding animals and people. As Jeri Laber and Barnett Rubin of Helsinki Watch summarized their findings, "the stories that we were to hear over and over again were these: 'The Russians bombed our village. Then the soldiers came. They killed women and children. They burned the wheat. They killed animals—cows, sheep, chickens. They took our food, put poison in the flour, stole our watches, jewelry, and money.'" The report continued, "The strategy of the Soviets and the Afghan government has been to spread terror in the countryside so that villagers will either be afraid to assist the resistance fighters who depend on them for food and shelter or will be forced to leave . . . We were told of brutal acts of violence by Soviet and Afghan forces: civilians burned alive, dynamited, beheaded; bound men forced to lie down on the road to be crushed by Soviet tanks; grenades thrown into rooms where women and children have been told to wait." Prisoners were summarily shot since the Russians claimed that they were illegal enemy combatants not covered by the Geneva Conventions.

The favored Russian weapons were anti-personnel mines. Although children particularly were attracted by plastic pens and red painted toy trucks, the effects on people could be reduced by clearing trails and inhabited areas. What really harmed the villagers was that the mines fell also on grazing areas so that livestock was crippled. As Malhuret continued, "When I arrived in Afghanistan for the first time in 1980, I was struck by the number of goats and cows that had legs in splints made of bamboo sticks and tied with wire . . . But the greatest loss, the herdsmen explained to me, is not so much the ones with splints, but rather all those animals that were killed from secondary infections."

The one major counterinsurgency tactic the Russians did not try was the relocation of people to internment camps as the British did in Malaysia and Kenya and the Americans did to "strategic hamlets" in Vietnam. In effect, Pakistan became their removal site, their strategic hamlet, as millions of Afghans fled the country under their relentless assaults.

When the last Russian troops crossed the Amu Darya river into Soviet territory in February 1989, at least one million Afghans had died. But that was not the end. A new force had been awakened by the Afghan war. As one *mujahid* told Peregrine Hodson with perhaps more foresight than he could have imagined, "The present war against the *Shuravi* is part of a greater war: the Islamic revolution. All over the world our brothers in the faith are awakening to a new spirit of religion." The Russians had planted dragon seed: they would keep on paying as they had for centuries in Chechnya and the Americans would begin to pay in Iraq. But even more than they, the Afghans would continue to pay as fighting continued among the warlords, then between the warlords and the Taliban, and finally between the Taliban and the Americans. There seems no end in sight.

CONCLUSION:

THE VERY EXPENSIVE SCHOOL

AS AMERICA IS LEARNING TO ITS GREAT AND GROWING COST, INSUR-gency is more than a topic of historical research. America is today embarked upon what appears to be a series of wars against insurgencies in the Middle East and Africa. So what emerges from a study of past insurgencies is of immediate and perhaps immense importance.

Consider the following: America is already deeply engaged in Afghanistan and Iraq and is fighting largely (but not solely) proxy wars in Somalia and the Philippines. Another campaign (as I write) is in the advanced planning and positioning stage against Iran, and still others are being discussed for various parts of Africa. At least one is being discussed for Latin America. The men who designed the current Bush Administration foreign policy, the neoconservatives, have called the combination of these campaigns the "long war" and have predicted—indeed proposed—that they will last about half a century.

In the spring of 2006, before he left office, Secretary of Defense Donald Rumsfeld approved three plans to fight the "long war" beyond Iraq and Afghanistan. Among other actions that have now been taken, the "Special Operations Command" has dispatched teams of Special Forces to at least twenty American embassies in Latin America, Africa, and the Middle East. These teams operate separately from the embassies and are not subject to control by the senior civilian American representatives, the ambassadors, as they engage in covert warfare not only against groups regarded as terrorist but even against states. This organization is now composed of fifty-three thousand men and operates on a budget in 2007 of $8 billion. Although these groups could bring America into war with any number of countries, they are treated by the Bush Administration as though not subject to congressional oversight or decision.

If this trend toward preemptive warfare is maintained and if what the neoconservatives, who are President Bush's principal advisers today, advocate continues to shape American policy toward military interventions or creates a momentum that the next administration, whether Republican or Democrat, will be unable to reverse, what will the effects be on American society and economy, on other nations, and on America's standing in the world community? These are issues I will now address on the basis of what has emerged from the historical record.

The campaign in Iraq, which has now lasted longer than the American involvement in the Second World War, has not been successful. New supposedly winning formulas have been announced periodically, but the only results have been daily casualty reports that remain painfully high for Americans and horrific for Iraqis. As of June 20, 2007, at least 3,521 American servicemen have lost their lives, and about twenty-five thousand have been wounded of whom half will never fully recover and many will spend the rest of their lives in hospitals. The human costs to Iraqis are nearly incalculable: perhaps as many as 601,027 (according to a study made by the Johns Hopkins School of Public Health) were killed between March 2003 and July 2006.

Almost two million Iraqis—about one in each twelve—have fled the country. That figure provided by the U.N. High Commissioner for Refugees, shocking though it is, understates the impact on Iraqi society because those who fled come almost exclusively from the Arab, and primarily from the Sunni, part of the population, which numbered only about seven million people. Thus, the more meaningful figure for those actually affected is about one in three or four. And in addition to these "external" refugees, more than one and a half million Iraqis have become "internal" refugees. That is, they have given up their houses, jobs, schools, and neighbors to seek security elsewhere in the country. If we consider this scale of misery in terms comparable to our own society, those figures equate to about thirty million Americans losing their homes and livelihood. Each month the numbers grow by about fifty thousand. And it gets worse. Even those who have not fled often live in deplorable conditions with massive overcrowding, unsanitary housing, and limited access to clean drinking water. In effect, the Sunni Arab population of Iraq has been decimated.

Some of the costs of the war cannot yet be reckoned. For the Americans, they include the following. The Department of Veterans Affairs has determined that up to October 2006 about one in five soldiers who served in Iraq has been "at least partly disabled," more than one hundred thousand have been granted disability compensation, and another one hundred thousand were expected to claim compensation. About two hundred thousand of the troops who have served in Iraq have suffered severe psychological damage or posttraumatic stress disorder (PTSD): in June 2007, a Pentagon mental health task force reported that 38 percent of soldiers, 31 percent of marines, and nearly half of the National Guard and reservists needed mental health treatment. Even that figure may be low because the numbers requesting help have so overwhelmed existing facilities that large numbers are being denied treatment.

Another fifty thousand or so have suffered concussions that will afflict them with memory loss, headaches, and confused thinking probably for the rest of their lives. Neurologists predict that hundreds of thousands more—at least one in each three soldiers who have engaged in combat for four months or longer—are in danger of blindness, deafness, or mental impairment. As one reporter put it, blast waves causing traumatic brain injuries "can leave a nineteen-year-old private who could easily run a six-minute mile unable to stand or even to think." Crassly put, they will not only be unable fully to contribute to American society but will be a burden on it. What this infirmity will mean in terms of their ability to function in their families and communities cannot be exactly determined, but the cumulative effects are bound to be severely detrimental.

Another number of Americans, on whom no statistics yet exist, will possibly develop cancer or will conceive children who will be born with severe defects because of exposure to the depleted uranium used in artillery shells and bombs. As General Brent Scowcroft remarked after the 1991 Gulf War, "Depleted uranium is more of a problem than we thought when it was developed." Recognition of that fact did not stop its use in the invasion of Iraq in 2003. Since many thousands of shells and bombs made of this heavy metal were used in both wars, the long-term effect on America (and Iraq) could be monumental.

The monetary costs of the war, while not so painful as the human costs, are enormous. Begin with America. So far the actual amount paid out or allocated for the invasion and occupation is approximately $500 billion. Current costs are running at more than $7.1 billion a month—$10 million an hour—and are rising more than 20 percent a year. Requested allocations in 2007, if approved, will take these figures far higher. The *direct* costs of the war are expected to rise to at least $700 billion, or nearly fifteen times as much as President Bush said the war would cost. But this outlay is only the tip of the iceberg. According to Nobel Prize laureate economist Joseph Stiglitz and former Assistant Secretary of Commerce Linda Bilmes, the *real* cost to America, as it would be figured by standard accounting methods, is between one and two *trillion* dollars. These figures are hard for most of us even to imagine. But they can be made concrete: about half that much would fund treatment for every American with heart disease or diabetes, double cancer research, and save millions of children's lives worldwide by providing immunization as well as really implementing the much hyped programs in public education here at home.

The costs to the people of Iraq have yet to be reckoned. In fact, taken individually, statistics are probably meaningless. The overall cost is simply the destruction of a whole society and a large part of the way it fed itself, housed itself, educated itself, kept itself in health.

What have these costs bought?

No well-informed observer believes that the war in Iraq is approaching success by any definition. The level of American troops has been between 130,000 and 160,000. Some supposed experts believe that double or triple that number would be required to "win." But, as the historical record makes clear, any increase in numbers is less likely to overawe the natives and produce "victory" than to provide more targets and increase their hostility to foreign rulers—us.

Despite targeted assassinations and fierce firefights, the Iraqi insurgency appears to be gaining rather than losing power. Even the most protected area in Iraq, the "Green Zone," which houses the Iraq government and the American personnel, was attacked in April 2007, and "U.S. officials warned that nowhere is safe in Baghdad" and the U.S. military commander warned that worse was to come. Whereas the insurgency originally was starved for funding, it now is self-financing. In fact, so flush with funds are insurgents that the price of weapons on the black market has surged at least a third in the last year. As I observed in Vietnam, many of the weapons are supplied by the Americans to the American-supported native army and police but end up in the hands of the insurgents.

What the invasion and occupation *have* accomplished is the destruction of the previous balances within Iraqi society. In the conditions of the war and the occupation, submerged ethnic hostilities have burst into a tragic civil war that may make reconstruction of the country impossible. So far, no new accommodation acceptable to the Iraqi people as a whole has been devised, much less accepted, so the future of the country is being debated with bombs and guns.

Perhaps inadvertently, America has fostered the civil war. The much-heralded 2005 elections were cast in terms not of policies or programs but ethnicity: they sharpened the divides between Kurds and Arabs and between Sunni Arabs and Shia Arabs. The resulting Shia-dominated government has used its new power, with the assistance of U.S. forces, to attack, drive away, or kill tens of thousands of Sunni Arabs. Police and army units created, trained, and paid for by the United States are the fist within the glove of the new Shia government. National reconciliation is not even a slogan.

It is so difficult to imagine how this conflict could be brought to a peaceful conclusion that some American observers think Iraq will break into at least three pieces. Some even advocate this breakup. Should it happen, the results will stretch far into the future and will be catastrophic for the Iraqis and even for the presumed objective of the Bush Administration to create a "new Middle East." The latest U.S. National Intelligence Estimate warns of the probability of a Turkish invasion of the Kurdish area of the north. While an invasion of the south, the presumably Shia area, is less likely, much closer relations with Iran, the growing power in the Middle East, have begun and are almost certain to increase. The relatively poor mainly Sunni section in the middle of Iraq will fester in resentment and nostalgia.

But, as leaders and supporters of the Bush Administration *rightly* say, "regardless of how we got here, we are here." Then they go on *wrongly* to draw the conclusion that we must "stay the course. What we are doing is working. It just needs more time, more troops, and more money." Translated into actual events and projected plans, what does this statement really mean?

On January 10, 2007, President Bush revealed a plan, already devised in early December 2006 by a leading neoconservative adviser, Frederick Kagan of the American Enterprise Institute, that would "totally Americanize the war in Iraq." The plan, "Choosing Victory: A Plan for Success in Iraq," would, as Terrell Arnold, former deputy director of counterterrorism and emergency planning and former chairman of the Department of International Studies at the National War College, has pointed out, put "almost all of American energy in

Iraq behind a military campaign to squelch Iraqi resistance." Yet the number of combat troops allocated to this task, if it were to meet the current concept of counterinsurgency, would have to be increased approximately sixfold.

Since no one believes a sixfold increase in American troops in Iraq is feasible, how can America engage in counterinsurgency there? The latest answer is given in the December 2006 *Counterinsurgency Field Manual* edited by Lieutenant General David H. Petraeus, USA, and Lieutenant General James F. Amos, USMC. That this is not just another study to be filed away in the Pentagon is shown by the fact that General Petraeus was appointed the U.S. forces commander in Iraq after President Bush fired his former commanders for opposing escalation. That is, he is being sent to implement the doctrine he and General Amos proclaimed. So what it says gives us a remarkable insight into what we can expect not only in Iraq but also elsewhere in the years to come. Here are some of the main points:

First, the purpose for issuing the manual: "With our soldiers and Marines fighting insurgents in Afghanistan and Iraq, it is essential that we give them a manual that provides principles and guidelines for counterinsurgency operations. Such guidance must be grounded in historical studies. However, it must be informed by contemporary experiences . . . Soldiers and Marines are expected to be nation builders as well as warriors."

Second, the role of the soldiers. No one, least of all the Iraqis, doubts that the American soldiers and marines are powerful warriors. But, as the manual makes clear, "Armed Forces cannot succeed in COIN [counterinsurgency] alone . . . killing insurgents—while necessary, especially with respect to extremists—by itself cannot defeat an insurgency . . . True extremists are unlikely to be reconciled to any other outcome than the one they seek; therefore, they must be killed or captured." However, as the manual notes, "killing every insurgent is normally impossible. Attempting to do so can also be counterproductive in some cases; it risks generating popular resentment, creating martyrs that motivate new recruits, and producing cycles of revenge." Although the manual touts its historical basis, the historical record shows precisely that killing insurgents is either impossible, irrelevant, or "counterproductive."

Third, "nation building." Have outsiders ever accomplished that task? Look at the American experience. American forces have been sent abroad to fight more than two hundred times since our country was founded. But in recent years only sixteen times have we attempted "the core objective of nation building . . . regime change or survivability." Of these sixteen, Minxin Pei and Sara Kasper found in a study for the Carnegie Endowment, eleven were "outright failures." Two, Germany and Japan, can be regarded as successes, while two others, tiny and nearby Grenada and Panama, were probably successful. Considering this record, John Tierney asked in the May 17, 2004, *International Herald Tribune,* how could neoconservatives or any conservatives "who normally do not trust their government to run a public school down the street, come to believe that federal bureaucrats could transform an entire nation in the alien culture of the Middle East?"

Fourth, legitimation. The manual states that "Political power is the central issue in insurgencies and counterinsurgencies; each side aims to get the people to accept its governance or authority as legitimate." Is this a feasible objective for foreigners? One searches the historical record in vain for an example of success. The foreign occupying force, by definition, is alien. Vietnam showed that even when the aliens had a numerous and established local ally, that ally (the South Vietnamese government) was more apt to be alienated by its association with the foreign military force than that force to be "nativized" by the natives. It is easy for both to become regarded as destructive to national purposes and the native culture. Nationalism, or more crudely xenophobia, overwhelms even beneficial programs. In the Spanish war against Napoleon, as we have seen, even when what the French wanted was to make the society more open, more productive, and more just, the Spanish people regarded those objectives as unimportant when weighed on the scales of nationalism. On what grounds can we expect that the attitude of Iraqis or Afghans or Somalis will be significantly different? Nowhere in the manual could this final problem be solved; indeed it could not even be adequately addressed. What is addressed is the means to separate the insurgents from the general population. But this was tried in Malaya (in fortified villages), Vietnam (with strategic hamlets), and Kenya (in detention camps). With the possible exception of Malaya (where the insurgents who as ethnic Chinese were themselves foreigners), these actions only further undermined any claim to legitimacy for the foreign occupiers. So, as we have seen in all the insurgencies described in the text of this book, *the single absolutely necessary ingredient in counterinsurgency is extremely unlikely ever to be available to foreigners.*

Fifth, adapting to the natives. Petraeus and Amos urge that the foreigners learn about the natives, even acquire some skill in their language, to minimize their foreignness. Yet refer again to the American Revolution: the British troops spoke the local language fluently, were generally of the same religion, and probably many were at least distantly related. The way these factors affected the American insurgency was exactly the opposite of what the British hoped: it caused the insurgents to try to subvert the British army rather than enabling the British to win over the insurgents. In practice, the American insurgents fought harder, more viciously, against their compatriots who sided with the British than against the British themselves. This was also true of the insurgencies in Spain, the Philippines, Yugoslavia, Greece, Kenya, Vietnam, and Afghanistan. It is true today in Iraq and Afghanistan.

Sixth, "Establish and expand secure areas." Like insurgency, counterinsurgency has a way of repeating itself. In Vietnam, in the 1960s, the Americans picked up and tried to apply the French *tache d'huile,* or enclave strategy. They hoped, as the U.S. Joint Chiefs of Staff then told *New York Times* correspondent Hanson Baldwin, "Once the enclaves were secure, the troops based in them could be employed either defensively or offensively, in small or large operations nearby or in other sections of South Vietnam." Very similar tactics have been and are being tried in both Iraq and Afghanistan: "Clear and hold" is presumably more sophisticated than "search and destroy." But as Mao Tse-tung and Vo Nguyen Giap among others have pointed out—and demonstrated—holding territory tends to weaken occupiers whereas it strengthens insurgents.

Finally, the use of force. What we end up with in Generals Petraeus's and Amos's manual is a series of technical ploys that, in the abstract, may be impressive but in the context of nationalism are peripheral to the issue as seen by the natives. And, as their best efforts are rejected, the Americans can be expected to react in anger against the ungrateful and stupid natives. Americans thus reacted in fury and incomprehension in the Philippines, Vietnam, the first Somalia campaign, Iraq, Afghanistan, and now Somalia again. Yet, as Petraeus and Amos point out, "The more force applied, the greater the chance of collateral damage and mistakes. Using substantial force also increases the opportunity for insurgent propaganda to portray lethal military activities as brutal." Using even limited and carefully targeted force seldom is more graciously accepted. We have seen this in the American Revolution when the foreign troops were extraordinarily disciplined, indeed even could be taken to a colonial American court if they tried to defend themselves against attacks by mobs or American terrorists. Their discipline did not make them any more acceptable to the Americans. It would be naïve to believe it to be different in Afghanistan and Iraq. Even the new senior American commander, Lieutenant General Raymond T. Odierno, believes that, under President Bush's new strategy, the war will last for years.

There is much more that could be said about the war in Iraq, but I will now turn to Afghanistan where a comparable but different war has been raging since the American invasion in October 2001. To put this in context, I will briefly recount what happened to Afghanistan following the Soviet withdrawal in February 1989.

When the Russians withdrew, the regime they had installed had enormous faults but it had one virtue: it was Afghan. Building on this virtue and the disunity of its opponents, it managed to retain power at least in the capital city, Kabul, for another four years. (That is, it survived almost exactly as long as the South Vietnamese government did after the beginning of the American withdrawal from Vietnam.) When it collapsed in 1993, the forces that had mobilized against the Russian occupation immediately turned against one another. The only thing on which they had agreed had been accomplished: the foreigners had been driven out.

The major beneficiaries of the $10 billion worth of arms supplied by the United States and Saudi Arabia were the half dozen "external" groups with offices in Pakistan. When their adherents began to reenter Afghanistan, they were more numerous, more unified, and better armed than the "internal" guerrillas who had done most of the fighting. (We have seen this same sequence of events in Algeria where the internal guerrillas were outnumbered about ten to one by external army.) Consequently, the hundreds of little groups of guerrillas who had fought the Russians found that they could not protect themselves or their property except by joining larger, more successful groups. Almost immediately, coalitions emerged under ambitious and ruthless men. Each "warlord," as these men came to be called, then attempted to seize enough of the country to become the ruler of all. Since all had been heavily armed, both by the Soviet Union, which had abandoned much of its weaponry, and by the United States, which for years had been supplying the Afghans with weapons to fight the Russians, the Afghans had the means to virtually destroy the country. They nearly did.

At this point, since expulsion of the Russians had been accomplished, the Americans and the Saudis no longer were willing to pour in money. The Afghans who had for nearly a generation fought and sacrificed were tired and hungry. But the warlords' need for money had, paradoxically, increased. No longer able to call on patriotism— or xenophobia—they had to buy support. They quickly found that only two sources of money were available: severe impositions on the native population (which often amounted to looting and the violence necessary to accomplish it) and promotion of Afghanistan's one saleable and highly lucrative product, opium. In the process of exploiting the people and engaging in the drug trade, the various warlord-led factions became what amounted to "megamafias." The only law was the gun, and the only security was membership in or protection by one of the factions.

As the able Pakistani journalist and frequent eyewitness Ahmed Rashid has written, "Afghanistan . . . was divided into warlord fiefdoms and all the warlords had fought, switched sides and fought again in a bewildering array of alliances, betrayals and bloodshed . . . The warlords seized homes and farms, threw out their occupants and handed them over to their supporters. The commanders abused the population at will, kidnapping young girls and boys for their sexual pleasure, robbing merchants in the bazaars and fighting and brawling in the streets." So painful to the general public had this condition become by 1996 that they welcomed the advent of a new force that promised to create a stable system of law and order and to chase away the men who had now actually become what the Russians always called them, bandits. This new force was the Taliban.

The Taliban have played—and are likely to continue to play— such a pivotal role in the Afghan insurgency that they demand understanding. The word *Taliban* in the language of the Pathan people, Pashtu, is the plural of the Arabic word *talib*, which means " a seeker [of truth], a student." It became the designation of the Afghan group because the leaders and most of their followers came out of religious seminaries, *madrasahs*, where they "sought the truth" as revealed in Islam. Like fundamentalists in Christianity, Judaism, and Hinduism, they believe their religious text to be the word of God. The Taliban are related to a number of other Muslim puritan revivalist movements that arose in the eighteenth and nineteenth centuries in reaction to Westeren intrusion. All shared the objectives of protecting their way of life from foreign ideas and their countries from imperialism. They thought they could gain the strength to do so if they purged their practices and their beliefs from what they regarded as corruption and returned to the basic concepts of their religion. In their religious motivation, they were comparable to the Puritans of New England and other Christian movements: all wanted to go back to fundamentals. The Arabic word for this "return" is *salafa*, so movements known as *salafiyah* spread throughout the Islamic world.

Today, fundamentalist movements are gaining popular support both because many Muslims see "the West" as the major danger to their independence and because many Africans and Asians have become disillusioned with what they have experienced of communism, socialism, capitalism, and democracy. Thus, Taliban-like movements are arising throughout the Muslim world.

As several of the subsequent leaders of the Taliban told Rashid, it was essentially the rapacity and corruption of the warlords that enabled the Taliban to win: the Taliban were encouraged by the merchant community whose livelihood was threatened, by common people who feared for their lives, and by Pakistan which viewed them as a tool of its foreign policy. They succeeded because there was no other force that promised peace and was acceptable to the Afghan population.

Like all Afghans, the Taliban were intensely religious. As I have mentioned, being a man of religion, a mullah, was virtually a national avocation. At least a quarter of a million men thought of themselves as mullahs in addition to engaging in whatever actually earned their living. Unlike most Afghans, the Taliban leaders were not just personally devout but were seminarians and teachers in religious schools. For them religion was a full-time calling. They had adopted this way of life because they were deeply motivated and doctrinaire.

Comparison with Spain in its war against Napoleon is suggestive. In both, the religious force was directed by thousands of active participants and was accepted by the general population; what distinguished Catholic Spain from Muslim Afghanistan was that Afghanistan had no central authority, no "church." That, in large part, is what the Taliban aspired to provide. But since Islam did not have a structure like the Church, the only unifying principle on which they could rely was a legal code. Theirs was drawn from primitive Islam and was mixed with Afghan tribal custom. So certain were they of the rightness of this code that they were determined to impose it upon the country.

For a while, the cost seemed worthwhile to many Afghans because, beginning in the southwestern province of Qandahar, the Taliban did indeed defeat the warlords, disarm the population, and restore trade. As Ahmed Rashid was told repeatedly in Afghanistan, the way the leader of the Taliban, Mohammed Omar, made the transition from mullah to militant was in response to the ugly tyranny of local Qandahari warlords. As the local myth put it, one of these men "had abducted two teenage girls, their heads had been shaved and they had been taken to a military camp and repeatedly raped. Omar enlisted some 30 *Talibs* who had only 16 rifles between them and attacked the base, freeing the girls and hanging the commander from the barrel of a tank." In a replay of the classic story of insurgency, they captured arms and ammunition with which they recruited more militants, and soon they were called to restore peace throughout their area.

Being relatively weak and untrained, the Taliban spread their control less with arms than with crafty politics and propaganda. Without exaggeration, one may say that the warlords dug their own graves. The Taliban managed to capture Kabul in September 1996; soon they fanned out over about 90 percent of the country. Initially they were welcomed because they brought security. Soon, however, the rigid, authoritarian, and often ugly order they imposed smothered the moves toward the better, freer life many had hoped would come when the Russians were chased away.

The principal crime of the Taliban in American eyes, however, was not the retrograde practices they imposed on their people but their espousal of Osama bin Ladin's jihad or religious war on America. The American government proclaimed that his movement, al-Qaida (El Qaeda as the media transliterates it), organized and carried out the September 11, 2001, bombings in New York and Washington. It is charged that members of the movement were also implicated in the 1998 attack on the American embassy in Nairobi and the suicide bombing of the American destroyer the USS *Cole* in Aden in October 2000. When the Taliban continued to protect al-Qaida, the United States launched an invasion on October 7, 2001. The attack was overwhelming: the Afghans had no defense against aerial attacks or modern weapons. About five thousand Afghans were killed and the Taliban government, simultaneously attacked by the remaining warlords, collapsed. The Americans were less successful against Bin

Ladin's group because al-Qaida is not a coherent group but a loose collection of people with similar beliefs and objectives.

American and British troops continued to chase after Bin Ladin and his allies, but what should be done about forming a new government of Afghanistan was less obvious. The American "ambassador," himself born an Afghan, called for the traditional national gathering, a sort of electoral college known as a *loya jirga,* to select a president. But, while the president it selected could be claimed to be legal, he had no power. Power was in the hands of the warlords and the foreign armies. Because the forty thousand foreign troops could not effectively administer the country—indeed, Afghans came to see them as another "infidel army trying to occupy their country"—the warlords did. They took up their former practices to such an extent that the Afghan people came to regard the Taliban with some nostalgia. As a village elder told Christian Parenti in October 2006, "If the Taliban come back, I will pray for them. I don't care if I have to grow a beard, go to mosque all the time. I don't care. At least they are not thieves." Increasingly, Afghans "are turning to Taliban-run courts, which are seen as more effective and fair than the corrupt official system . . . people throughout the country, including in Kabul, report that crime is increasing—and complain that the police are the main criminals."

Actions by the British and American troops appear, predictably, to worsen the situation. The British defense secretary admitted in July 2006 that "the deployment of 3,300 British forces . . . has energized the Taliban." Other witnesses tell why: *New York Times* reporter Carlotta Gall drove through one district where she saw that "to fight their way into this area and clear it of Taliban insurgents, NATO troops bulldozed through orchards, smashed down walls and even houses, and churned vineyards and melon fields to dust . . . 'They did not come to bring peace for us, they came to destroy us,' said Hajji Abdul Ghafar, 60, an elder on the Sperwan village council . . ." As another reporter found, "hatred and mistrust of America is bitter and intense . . . New enemies . . . seem born every day."

It isn't only the actions of the military that create hatred. It is also popular despair. Kabul is a city in ruins: there is little drinking water and most of that is unsanitary; electricity is available only a few hours a day; there is no health care system except the little that foreign non-governmental organizations can provide; schools are without teachers; unemployment is ruinous. Even educated women, hired by the American-controlled government, are so ill-paid that they are forced into prostitution to support their children. Government is viewed as little more than a criminal organization.

Afghanistan's only efficient industry is the drug trade. In 2006 Afghanistan produced about 90 percent of the world's opium. Ironically, the only effective deterrent to the trade is Iran, which has deployed a force almost as large as the American army in Afghanistan and built barriers against smugglers more impressive than the wall America constructed along the Mexican frontier to try to interdict illegal immigration. Battling the virtual armies of the drug lords, Iran has lost about as many police and soldiers as America has lost in Iraq—thirty-five hundred killed and fifteen thousand wounded. Neither they nor the Americans have been able to stop the cross-border smugglers. Of course, a large part of the reason is that Westerners are willing to buy drugs. Within Afghanistan, poor people raise it because they have no other way to feed themselves. Thus, when American and British soldiers burn the farmers' fields of poppies, even this sordid aspect of the insurgency has become a popular, national issue. The Taliban, who initially forbade and nearly wiped out the drug trade, now take the farmers' side.

As the *New York Times* editorialized, "The Taliban are enjoying a resurgence in presence and power . . . the American-backed government is in danger of losing the battle for Afghan hearts and minds. If this battle is lost, there can be no lasting military success against the Taliban and their Qaeda allies." Even the then senior Republican in the Senate Bill Frist, so despaired that he suggested that some Taliban be brought back into the government. It may be too late even for this ploy: the Taliban are gaining power almost daily despite the overwhelming military power of America and Britain. Some see a replay of the Soviet defeat.

*　　*　　*

Relative to Iraq and Afghanistan, Somalia is a small country and so far a small problem for America. What little most Americans know about Somalia comes from the literally explosive film *Black Hawk Down*. If we leave aside the human interest—our decent young men caught in an unwinnable fight against evil men in a faraway country—the two messages of the film are set out simply: at the beginning we see the poor people of the country being starved and murdered by a gangster regime, and at the end we see that, despite the horror they have suffered at the hands of their native warlords, the Somalis do not want foreigners, Americans, to rule their lives.

Faced with unacceptable casualties, the Clinton Administration pulled out and left the Somalis to their own devices. Those devices were perhaps even worse than in Afghanistan. Warlords pillaged, raped, and killed. So horrible had life become that eventually a group arose within the country to chase out the gangsters. What motivated the group was religious fundamentalism. Having long ago become disillusioned with secular nationalism and Marxism, as have most of the peoples of the Middle East, Africa, and Asia, a number of Somalis turned to Islam to achieve a sufficient degree of unity to recapture their country. They created a form of Islamic government that they called the "Islamic Courts Council," as the Afghans had done with the Taliban, emphasizing Islamic law to create domestic tranquillity. As one observer put it, "during their 6-month reign [they] pacified the hornet's net of Mogadishu without foreign peacekeepers or significant foreign aid. They succeeded by getting clans to voluntarily disarm their militias and getting Somalis to buy into their Islam-is-the answer solution." That commitment put the Somalis athwart American policy.

At that point, the Bush Administration, fearing that the religious establishment was part of what it sees as a widespread Islamic terrorist organization, al-Qaida, encouraged neighboring Ethiopia, which has a Christian government, to attack Somalia. America provided the arms and money. Ethiopia, which is widely believed in the Horn of Africa to have long aimed to dominate Somalia, sent its forces in and captured Mogadishu. The Muslim leaders attempted to flee to Kenya but were intercepted by American aircraft and Special Forces at the frontier.

Many were killed. The problem then arose, who could be installed as the new government? Back to *Black Hawk Down:* the same people who we were then horrified to see raping and pillaging the country, the warlords have returned with American encouragement, arms, and money astride their machine gun–decked "technicals." And, just as they were portrayed doing in *Black Hawk Down,* gangs under the warlords have created "a gigantic protection racket . . . restricting food aid deliveries to hundreds of thousands of civilians, who are also being charged to shelter under trees . . . [while being subjected to] systematic looting, extortion, and rape by uniformed troops."

As the great German philosopher of history Georg Wilhelm Hegel warned us, "peoples and governments never have learned anything from history or acted on principles deduced from it." As though to prove Hegel's point, the senior American official for African affairs, Jendayi Frazer, pontificated, "No insurgency group can survive without support from neighboring countries." In fact, as I have demonstrated in this book, most insurgencies have done so successfully. Moreover, as the senior CIA officer dealing with the anti-Soviet war in Afghanistan observed, ". . . in the last fifty years, no nationalist-based insurgency against a foreign occupation has lost . . ." Yet, this lack of knowledge of history has apparently led the Pentagon to use the Somali campaign as a blueprint or military model—using groups like the one deployed in Somalia, "a secret American Special Operations Unit, Task Force 88," and AC-130 gunships to capture or kill Muslim insurgents—for missions in other parts of Africa, Asia, and Latin America under authority granted it by the Bush Administration.

What will result is predictable. As an able local observer, Salim Lone, a journalist in neighboring Kenya and a former member of the UN mission to Iraq, wrote, "An insurgency by Somalis, millions of whom live in Kenya and Ethiopia, will surely ensue, and attract thousands of new anti-U.S militants and terrorists." Like Iraq and Afghanistan, Somalia has plunged into a civil war in which the society is collapsing.

*　　*　　*

Stretching out ahead of these costly, long-lasting, and unwinnable wars is what the neoconservatives have named the "long war." To fight it, the Bush Administration has called for an immediate increase in the American army by almost a hundred thousand men and women and has built an astonishing array of *737 military and air bases* throughout the world, with particular emphasis on Eastern Europe, Central Asia, and various parts of Africa. The next likely stage, being threatened by the Bush Administration at the time of this writing, is an attack on Iran.

To begin that fight, the Bush Administration has positioned carrier task forces of the U.S. Navy adjacent to places it has proclaimed to be likely targets. Two such powerful task forces are in or near the Persian Gulf on the frontier of Iran, and another is operating in the Indian Ocean off the coast of Somalia.

What would happen if America or Israel does attack Iran can be predicted: Iran is much larger than Iraq and has a population about three times the size. Moreover, it has learned from the Iraq war: apart from its 850,000-man regular army, it has long since been preparing for protracted guerrilla war both on land with its 150,000-man *Pasdaran-i Inqilab* (Revolutionary National Guard) and at sea with its fleet of speedboats operating on the gulf. It has a relatively sophisticated, middle-range missile armory. More significant, it has overseas assets that will enable it (or its fifteen million fellow Shiis who control the Iraqi government) to carry the war to American forces in Iraq and its battle-tested Lebanese Hizbullah allies to carry the war to America's ally, Israel. Since it produces about 5 percent of the world's energy, disruption of that source would impact worldwide prices; even more important, since 40 percent of the world's energy ships through the gulf, any disruption in an American invasion could send oil prices skyrocketing—every $1 rise in the price of oil costs the United States more than $3 billion in national income. The anticipated rise in price in an American attack on Iran would plunge the world into depression.

Unfortunately, this course of American militarism is not just speculation. It has long been advocated by the neoconservatives and is now at the top of the Bush Administration agenda. And it is only the next step of a long-term strategy laid out by neoconservative James Woolsley for a generational conflict—lasting perhaps forty years—all over the world. As the conservative English journal *The Economist* editorialized, "the Neoconservatives are not conservatives. They are radicals. Their agenda adds up to a world-wide crusade. With all its historic, anti-Muslim connotations, it is precisely the word most calculated to perpetuate movement down the path desired by the Neoconservatives, permanent, unending war."

The way to bring about this nightmare was laid out in the March 2005 "National Defense Strategy of the United States of America," which Defense Secretary Donald Rumsfeld approved before he left office. It calls for the United States to maintain its ability

> to operate in and from the global commons-space, international waters and airspace, and cyberspace . . . Such capacity provides our forces operational freedom of action . . . Our role in the world depends on effectively projecting and sustaining our forces in distant environments where adversaries may seek to deny us access . . . A key goal is developing the capability to surge military forces rapidly from strategic distances to deny adversaries sanctuary . . . *Irregular* conflict [that is, insurgency] will be a key challenge for the foreseeable future. [Since the areas of military action span the globe,] the new global posture—using main operating bases (MOB), forward operating sites (FOS), and a diverse array of more austere cooperative security locations (CSL)—will support such needs. In addition, our prepositioned equipment and stocks overseas will be better configured and positioned for global employment.

To enable this to happen, the strategy paper called for "new legal arrangements that maximize our freedom to: Deploy our forces as needed; conduct essential training with partners in the host nation; and support deployed forces around the world." It also called for arrangements that would prevent the "transfers of U.S. personnel to the International Criminal Court," thus effectively extra-territorializing Americans engaged in counterinsurgency everywhere no matter what actions they might take or what crimes they might commit.

"Campaigns to 'win decisively' are undertaken to bring about fundamental, favorable change in a crisis region and create enduring results. They may entail lengthy periods of both major combat and stability operations; require regime change, defense, or restoration; and entail significant investments of the nation's resources and time. 'Win decisive' campaigns . . . include Operations Just Cause and Iraqi Freedom." Somalia follows Iraq and Afghanistan as the next step in this forty-year war.

Apart from what happens abroad, the war has a frightening domestic impact. The monetary cost of such a war has been estimated to be at least $15 *trillion*. But that is, relatively speaking, the trivial part of the cost. The real costs are both personal and institutional: on the personal level, combat invariably promotes violence, and overwhelming power combined with fear promotes what might be termed casual violence. In a study published by the Pentagon on May 4, 2007, the Surgeon General found a significant decline in moral standards among American soldiers and marines. About one in each ten reported mistreating noncombatants and needlessly destroying their property, and two in three said they would not report such illegal actions. On the institutional level, prolonged combat—and already the wars in Iraq and Afghanistan have lasted longer than the Second World War—could bring about the destruction of the world in which we live and the replacement of our civic, cultural, and material "good life" by something like the nightmare George Orwell predicted in his novel *1984*. As the former editor of the *New York Times* Joseph Lelyveld has written, the Bush Administration "came close to asserting the power of the commander in chief to declare anyone in the world, of whatever citizenship or location, 'an unlawful enemy combatant' and—solely on the basis of that designation—to detain the person indefinitely without charge, beyond the reach of any court. As we now know, it then acted on its own theory; according to a list being compiled by Human Rights Watch, alleged terrorists were detained at American behest in Mauritania, Bosnia, Indonesia, the United Arab Emirates, and Yemen—as well as Afghanistan and the border areas of Pakistan where most al-Qaeda and Taliban fighters were captured." Even if, indeed especially if, such violations of the fundamental rights enshrined in our Constitution are justified by the "Global War on Terror," these actions make the case that this policy is moving us toward the destruction of the very values it is alleged to be designed to protect.

This, in short, is not just rhetoric or speculation: it is made up of operational plans, dedicated military personnel, operating from existing bases, with already constructed and positioned weapons, and sustained by an already allocated budget. It has already been approved by the Bush Administration. Is it to be our future? Former President Dwight Eisenhower gave us one answer when he said:

"Every gun that is made, every warship launched, every rocket fired signifies, in the final sense, a theft from those who hunger and are not fed, those who are cold and are not clothed. The world in arms is not spending money alone. It is spending the sweat of its laborers, the genius of its scientists, the hopes of its children . . . This is not a way of life at all, in any true sense. Under the cloud of threatening war, it is humanity hanging from a cross of iron."

NOTES

xvii   *No More Vietnams?* They included, in addition to Mr. Pfeffer and myself, Frances FitzGerald, Eqbal Ahmad, Richard J. Barnet, Chester Cooper, Theodore Draper, Luigi Einaudi, Daniel Ellsberg, John King Fairbank, Colonel (later Lieutenant General) Fred Haynes, USMC, Stanley Hoffman, Samuel P. Huntington, George Kahin, Henry Kissinger, John McDermott, Hans J. Morgenthau, Ithiel de Sola Pool, Edwin Reischauer, John Reilly, Arthur Schlesinger, Jr., Sir Robert Thompson, James C. Thomson, Jr., Leroy Wehrle, Albert Wohlstetter, and Adam Yarmolinsky.

xix   *Cetniks* The C in Cetniks is pronounced "ch," so the word is sometimes written Chetniks or Çetniks.

xxii   *George Carver wrote* George Carver's article in *Foreign Affairs*: *Pentagon Papers*, I/334–335.

xxii   *Bernard Fall* The *Pentagon Papers* ("Senator Gravel Edition," Boston: Beacon Press, 1971), I/336.

xxii   *oil spot* *Principes de pacification et d'organisation,* quoted in Joseph Buttinger, *Vietnam: A Dragon Embattled* (New York: Praeger, 2 vols. 1967), 479.

xxiii   *"failed dismally"* *Pentagon Papers,* II/128.

xxiv   *"on a blotter"* Robert Taber, *War of the Flea: The Classic Study of Guerrilla Warfare* (New York: Stuart, 1965, reprinted Dulles, Va.: Potomac Books, 2002), 49–50.

xxv   *Mao Tse-tung* *Yu Chi Chan,* translated by Brigadier General Samuel B. Griffith, USMC, as *Mao Tse-tung on Guerrilla Warfare* (New York: Praeger, 1961), 94ff., esp. 113.

2   *"the attack ceases"* In a memorandum entitled "Indian Warfare" which was reprinted in Lewis Butler, *The Annals of the King's Royal Rifle Corps,* (London, 1913), I, 159–160 and quoted in Eric Robson, "The Armed Forces and the Art of War," in J. O. Lindsay (ed.), *The New Cambridge*

Modern History, vol. 7, *The Old Regime, 1713–63* (Cambridge: Cambridge University Press, 1970), 173.

4   *to dig trenches*   John Grenier, *The First Way of War* (Cambridge: Cambridge University Press, 2005), 87–88.

4   *for things British*   On Washington's personal tastes and Benjamin Franklin's comments on American admiration for English goods, see William R. Polk, *The Birth of America* (New York: HarperCollins, 2006), 148.

4   *American "Continental Line"*   Despite its initial disapproval of a standing army, Congress in May 1778 approved the British regimental system for the Continental Line. Robert K. Wright, Jr., *The Continental Army* (Washington, D.C.: Center of Military History, United States Army, 1989), 124–126. On the role of Frederick von Steuben in 1778 at Valley Forge, see pages 140ff. Steuben's *Regulations for the Order and Discipline of the Troops of the United States* was printed in 1779 at Philadelphia and passed out to officers throughout the Continental Line.

4   *"by siege and assault"*   Charles Royster, *A Revolutionary People at War: The Continental Army and American Character, 1775–1783* (New York: Norton/Chapel Hill: University of North Carolina Press, 1979), 116.

5   *"had not been well founded"*   John Bigelow (ed.), *The Life of Benjamin Franklin Written by Himself* (Oxford: Geoffrey Cumberlege, 1924), 189–190.

5   *into such a strategy*   Like virtually everything on the Revolution, this question has provoked a vast array of books. See, for example, Pauline Maier, *From Resistance to Revolution* (New York: Vintage, 1974), passim; Dirk Hoerder, *Crowd Action in Revolutionary Massachusetts 1765–1780* (New York: Academic Press, 1977), Part II; and Polk, *Birth of America*, 238ff.

6   *"any foreign market"*   Leonard Labaree (ed.), *The Papers of Benjamin Franklin* (New Haven: Yale University Press, 1968), 12, 183–189.

6   *made him a patriot*   J. H. Plumb, *England in the Eighteenth Century, 1714–1815* (Harmondsworth: Penguin, 1969), 126, and A. M. Schlesinger, *The Colonial Merchants and the American Revolution* (New York: Columbia University Press, 1918), quoted in Esmond Wright, "American Independence in Its American Context" in A. Goodwin (ed.), *The New Cambridge Modern History* (Cambridge: Cambridge University Press, 1971), 8, 522.

6   *extensively studied*   Hiller B. Zobel, *The Boston Massacre* (New York: Norton, 1970) is the classic.

7   *sale of tax stamps*   A letter written on the day following the Stamp Act Riot, November 2, 1765, Henry B. Dawson, *New York City During the American Revolution, Being a Collection of Original Papers from the Manuscripts in the Possession of the Mercantile Library Association of New York City* (New York: privately printed, 1861).

7   *"extreme hazard and suffering"*   "The Military Conflict Considered as a Revolutionary War," in Stephen G. Kurtz and James H. Hutson (eds.),

Essays on the American Revolution (Chapel Hill: University of North Carolina Press, 1973); reprinted in John W. Shy, *A People Numerous and Armed* (New York: Oxford University Press, 1976), 215.

7   *two were Loyalists*   Some estimates are higher. Paul Smith, "The American Loyalists," *The William and Mary Quarterly* 25, no. 2 (April 1968).

10   *"the civilian population"*   John Pancake, *1777: The Year of the Hangman* (University: University of Alabama Press, 1977), 83.

11   *"conducted by military means"*   "Kurtz and Hutson, quoted in Pancake, 84.

11   *"made in the country"*   Amos Kendall, *The Life of Andrew Jackson* (New York: Harper, 1843), 19–20, quoted in Blackwell Robinson, *William R. Davie* (Chapel Hill: University of North Carolina Press, 1957), 40.

12   *to effect it*   In *Polk's Folly* (New York: Doubleday, 2000) I describe the work of one group led by Colonel Thomas Polk in Mecklenburg, North Carolina, 114ff.

12   *often with terrorism*   Pancake, 102.

12   *As Catherine S. Crary writes*   *The Price of Loyalty: Tory Writings from the Revolutionary Era* (New York: McGraw-Hill, 1973), 7.

13   *"burning their crops"*   *This Destructive War: The British Campaign in the Carolinas, 1780–1782* (Tuscaloosa: University of Alabama Press, 1985), 25.

13   *encouraged to come forth*   Pancake, 33.

13   *"on that river"*   Pancake, 98.

13   *stretched out in Vietnam*   Another part of this experience was to be repeated in the American occupation of Iraq. Cornwallis was forced to emphasize the role of the Loyalists because, in London, the government was being told by Loyalist émigrés that the Americans were devoted to England and with just a bit more help would come out in great numbers to support the Royal cause. Governments, then as now, often listen to the siren song of exiles instead of their men on the ground.

13   *One Scottish visitor describes*   Janet Schaw quoted in Crary, 61. The informal war on Loyalists in the Boston area, led by Samuel Adams, is recounted particularly well by Zobel, *Boston Massacre*; Hoerder, *Crowd Action in Revolutionary Massachusetts*; and Bernard Bailyn, *The Ordeal of Thomas Hutchinson* (Cambridge: Harvard University Press, 1976).

15   *As John Shy writes*   Shy, 142.

16   *From personal experience*   Born in England in 1731, Lee was commissioned in the British army and like George Washington fought under General Braddock in the French and Indian war; he rose to the rank of lieutenant colonel during the British campaign in Spain, where he distinguished himself as a daring and brave officer. During service in Poland, where he first observed partisan warfare, he was commissioned a major general. He moved to America in 1773. When the Revolution broke out, he was commissioned a major general at the urging of George Washington.

16   *put it bluntly*   Writing in the *Virginia Gazette* of July 29, 1775, quoted in Charles Royster, 26–27.

16   *"can alone Succeed"*   "A Plan for the Formation of the American Army . . . [April 1778]," *Lee Papers*, II, 383–389, quoted in Shy, 154–155.

16   *"the mighty British empire"*   In my book *The Birth of America* (New York: HarperCollins, 2006), 298–309, I show how I believe the British government evaluated the conflict with the colonies.

17   *"lower class of these people"*   Quoted in Shy, 146.

18   *the Carolinas and Georgia*   The first such battle was known as the battle of Moore's Creek Bridge on February 27, 1776, and one of the last was at The Cowpens in 1780. On taking up his command, the outstanding regular officer, Washington's protégé General Nathanael Greene, thought that these militiamen were "the worst in the world." Charles Patrick Neimeyer, *America Goes to War* (New York: New York University Press, 1996), 9. See Robert D. Bass, *Swamp Fox: The Life and Campaigns of General Francis Marion* (New York: Holt, 1959) and Chalmers Gaston Davidson, *Piedmont Partisan* (Davidson, N.C.: Davidson College, 1951).

20   *urban populations dwindling*   Raymond Rudforff, *War to the Death: The Sieges of Saragossa 1808–1809* (London: Hamish Hamilton Book Club Edition by Purnell Book Services, 1974), 5.

20   *"look at what you have become"*   Reproduced in W. N. Hargreaves-Mawdsley, *Spain under the Bourbons, 1700–1833* (Columbia: University of South Carolina Press, 1973), 211–212.

21   *As Raymond Rudforff wrote*   Rudforff, 6.

21   *the mistress of a commoner*   His common origin was memorialized in his popular nickname, *el Choricero* (the "sausage maker").

22   *"came from France"*   *The Structures of Everyday Life, Volume I of Civilization and Capitalism, 15th–18th century* (London: Collins, 1981), 54–55.

23   *as a wife*   Reproduced in Hargreaves-Mawdsley, 196–197. The king's proclamation is on 199 and his son's apology on page 200.

23   *"law into their own hands"*   Reproduced in Hargreaves-Mawdsley, 204–205.

24   *"between abdication and death"*   Pedro Cervallos, *Exposición de los hechos ye máquinaciones que had preparado la usurpación de la corona de España y los medios que el Emperador de los franceses ha puesto en obra para realizarla* (Cádiz, 1808), 30, quoted in John Lawrence Tone, *The Fatal Knot: The Guerrilla War in Navarre and the Defeat of Napoleon in Spain* (Chapel Hill: University of North Carolina Press, 1994), 48.

24   *to be burned down*   Orders of the day, published in both Spanish and French, on May 4, 1808, reproduced in Hargreaves-Mawdsley, 206–207.

25   *were even illiterate*   Rudforff, 44.

26   *"killing the French heretical dogs"*   Rudforff, 42–43.

26   *"would be reborn three days later in paradise"*   Tone, 60–61.

26   *defeat French armies in battle*   They actually forced a French army sur-
      render at Bailén when the French made a tactical mistake.

27   *The guerrilleros*   In modern usage, the word *guerrilleros* has been replaced
      by guerrillas. I shall use it in that form.

27   *"tortured to death by a flea"*   H. von Brandt, *The Two Minas and the Span-
      ish Guerrillas* (London, 1825), quoted in Charles Esdaile, "'Heroes or
      Villains' Revisited: Fresh Thought on *La Guerrilla*," in Ian Fletcher (ed.),
      The Peninsula War (Staplehurst, England: Spellmount, 1998), 95.

27   *French cavalry and artillery*   Tone, 10.

28   *revenge for offense*   Still the best study is Julian Pitt-Rivers, *The People
      of the Sierra* (Chicago: University of Chicago Press, 1971); also see his
      introduction to *Mediterranean Countrymen* (Paris: Mouton, 1963).

29   *in some 111 bands*   Public Record Office, WO I/40 cited in René Char-
      trand, *Spanish Guerrillas in the Peninsular War 1808–1814* (London: Os-
      prey, 2004), 17.

29   *about thirty-five thousand combatants in total*   Chartrand, 17.

31   *we call counterinsurgency*   Much of the following information is drawn
      from Tone, which is the best account I have found of this episode of
      Spanish history. Particularly see chap. 5.

31   *the first of many failures of counterinsurgency*   Tone, 65–66.

32   *mobilization of civilians*   A summary of the twenty-four articles is given
      in Miguel Angel Martín Mas, *The Guerrilla Wars, 1808–1814* (Madrid:
      Andrea Press, 2005), 14. They were supplemented by a new decree in the
      name of Fernando in April 1809.

32   *into the Spanish army*   Esdaile, 97. That is, to drop *guerrilla* and take
      up *guerra*. This was always the temptation as guerrilla leaders from the
      American insurgency to Vo Nguyen Giap in Vietnam demonstrate.

33   *into outlaws*   Many of the leaders were exiled, imprisoned, or executed
      during the Restoration. When I discuss the IRA in Ireland, I will show
      that a similar policy was begun under the newly independent government
      of Éamon de Valera and in Algeria under Ahmed Ben Bella in suppress-
      ing the "internal" army that had fought the French. The guerrillas had
      outlived their usefulness.

33   *the most wide-ranging military career of all time*   Of Swiss origin, he had
      joined the French army in 1798. After a period out of service, during
      which time he wrote the first of his several books on strategy, he was
      commissioned by Napoleon and took part in several of Napoleon's major
      European battles. Then he was sent to Spain, where he closely observed
      the guerrilla war. Disappointed in promotion in the French army, he
      joined the Russian army, where he became a lieutenant general. In Rus-
      sian service, he took part in campaigns against the Ottoman Empire and
      helped Tsar Nicholas plan the defense of Sevastopol during the Crimean
      War. In his final incarnation, he helped Emperor Napoleon III plan his
      Italian expedition in 1859. His most famous work on strategy is his 1838

Précis de l'art de la guerre, which I quote in the 1862 English translation in the West Point Military Library, where it was used as a textbook for cadets.

38   *not Spanish but Tagalog: Katipunan*   The *Kataastaasang Kagalang-galan-gang Katipunan ng mga anak ng Inang Bayan.* "The Highest and Most Noble Society *(Katipunan)* of the Sons of the Nation."

39   *with a purse full of gold*   Allegedly, he was bribed with the equivalent of $800,000, as Richard Kessler recounts in *Rebellion and Repression in the Philippines* (New Haven: Yale University Press, 1989), 10.

40   *the same as the Cubans will be"*   Leon Wolff, *Little Brown Brother: America's Forgotten Bid for Empire Which Cost 250,000 Lives* (London: Longmans, 1961), 45–46.

41   *"and slept soundly"*   Leon Wolff, 174.

42   *"burning their dwellings"*   I have used the third edition (London: HMSO, 1906) which was reprinted in photo offset with an introduction by Douglas Porch, professor of strategy at the Naval War College (Lincoln: University of Nebraska Press, 1996), 128–129, 132, 145.

42   *the natives were "gugus"*   Stuart Creighton Miller, "The American Soldier and the Conquest of the Philippines," in Peter W. Stanley (ed.), *Reappraising and Empire: New Perspectives on Philippine-American History* (Cambridge: Harvard University Press, 1984), 31–32. The American military governor requested that black American soldiers not be sent out and noted that white soldiers refused to salute black officers and sometimes jeered them, shouting "what are you coons doing here?" A few were so angry that they deserted to the Filipinos, but overall black regiments served loyally and well.

42   *the politics of insurgency*   Aguinaldo appears to have had a personally rather checkered career. Having been bought off by the Spaniards, he worked closely with Admiral Dewey during the American invasion. Dewey makes this clear in testimony before Congress that is reprinted in Henry F. Graff, *American Imperialism and the Philippine Insurrection* (Boston: Little Brown, 1969). Only when let down by Dewey did he emerge as a true nationalist.

44   *was not to be reported*   As Leon Wolff wryly commented, in a foretaste of Vietnam nearly seventy years later, eleven journalists retreated to the British Crown Colony Hong Kong where they could evade censorship to warn their readers that "owing to official dispatches from Manila made public in Washington, the people of the United States have not received a correct impression of the situation in the Philippines . . ." Reprinted in Wolff, 262.

44   *at least temporarily true*   Wolff, 288–289.

46   *"to send detachments"*   Benedict J. Kerkvliet, *The Huk Rebellion: A Study of Peasant Revolt in the Philippines* (Berkeley: University of California, 1977), 54.

47	*during the occupation* General Edward Lansdale, *In the Midst of Wars: An American's Mission to Southeast Asia* (New York: Harper & Row, 1972), 8.
47	*"food to the Huk soldiers"* Kerkvliet, 63–65; the quotation is from page 67.
48	*American army put it, "furiously"* Official history of the U.S. Sixth Army. To the contrary, the USAFFE-sponsored guerrillas were said to be a "passive intelligence operation." Kerkvliet, 93–94. The Filipino who turned out to be the most important member of the group for future events, the later president, Ramón Magsaysay, did not arise from within the guerrilla movement but was appointed to a regional leadership position by American military officers. Lansdale, 34.
48	*recognizing Philippine independence* Or it nearly did. There were strings attached. The Philippine government granted the United States a ninety-nine-year lease on some twenty-three military bases including the giant Clark airfield and the Subic Bay naval base. These grants were to rankle until they were given up in 1992.
49	*As Richard Kessler* commented Kessler, 32.
49	*for absentee owners* Only a few thousand in the total population then thought to be about twenty-one million.
49	*a much larger group of supporters* Colonel Napoleon D. Valeriano and Lieutenant Colonel Charles T. R. Bohannan, *Counterguerrilla Operations: The Philippine Experience* (New York: Praeger, 1962), 23.
50	*appointed by the Americans* He was appointed by Colonel Gyles Merrill of the U.S. Army to be commander of one of the units sponsored by Squadron 311, and later the U.S. Army made him military governor of a province. In 1946 he was elected to the Philippine House of Representatives. Lansdale, 34.
50	*an American "007"—James Bond* Colonel Lansdale was famously portrayed as "Colonel Hillandale" in the William J. Lederer and Eugene Burdick novel *The Ugly American.*
50	*more effectively both day and night* As he wrote, "I had been getting along with little sleep. With my new roommate, I got even less. Each night we sat up late discussing the current situation. Magsaysay would air his views. Afterwards, I would sort them out aloud for him while underscoring the principles of strategy or tactics involved." Lansdale, 37.
51	*Lansdale minutely reports* Lansdale, 28.
51	*"was rotten with corruption"* Lansdale, 25.
52	*were said to have surrendered* Lansdale, 50–51.
52	*"land for the landless"* Lansdale, 52.
52	*from abroad to their families* Since these programs were set in motion, they mounted year by year so that in 2004, they aggregated about $15 billion yearly and almost one in every ten Filipinos was working overseas.
53	*"malaise that the dirty war represents will remain"* *International Herald Tribune,* August 22, 2006.

55 *faded into the Irish population* J. C. Beckett, *The Making of Modern Ireland, 1603–1923* (New York: Knopf, 1966), 14.

56 *"the end of their feast"* *Annals of the Four Masters* (Bagwell, Ireland, 1885), II/289.

56 *the military campaigns proved ineffective* I have briefly touched on these events in *Polk's Folly* (New York: Doubleday, 2000 and Anchor, 2001), 16ff. More information is given in T. W. Moody, F. X. Martin, and F. J. Byrne (eds.), *A New History of Ireland* (Oxford: Oxford University Press, vol. 3, 1976); Brendan Fitzpatrick, *Seventeenth Century Ireland: The War of Religions* (Dublin: Gill & Macmillan, 1988); and Jonathan Bardon, *A History of Ulster* (Belfast: Blackstaff Press, 1992).

57 *incoming English and Scots colonists* Nicholas Canny, *The Elizabethan Conquest of Ireland: A Pattern Established, 1565–1576* (Hassocks, Sussex: Harvester Press, 1976).

57 *aim was to break Ireland's cultural tradition* In essence, this was the policy used by the Spaniards in the New World and the Philippines against the natives; there, of course, it was Spanish and Catholicism that were the tools of colonialism, whereas in Ireland it was English and Protestantism.

57 *service to the British Crown* *English Society in the Eighteenth Century* (London: Penguin Books, 1991), 34–35.

58 *immigrated to America* Kerby A. Miller, *Emigrants and Exiles* (New York: Oxford University Press, 1985), 280.

59 *"demolishing of tenants' homes"* Miller, 288.

59 *"no Irish Republican Army, no I.R.A."* Tim Pat Coogan, *The IRA, A History* (Niwot, Colo.: Roberts Rinehart, 1993), 4.

61 *some 2,000 men resigned* M. L. R. Smith, *Fighting for Ireland?* (London: Routledge, 1995), 34. When he wrote this account Mr. Smith was on the staff of the Royal Naval College.

61 *almost automatic among young Irishmen* This and the subsequent suppression of the militants is the subject of a remarkable film directed by Ken Loach entitled *The Wind That Shakes the Barley*, which won the Palme d'Or at the 2006 Cannes Film Festival.

62 *would not tolerate an expensive new war* This was the major motivation behind the 1921 Cairo Conference that the then colonial secretary Winston Churchill organized to settle the status of Britain's colonies and territories in the Middle East. On it, see William R. Polk, *The United States and the Arab World* (Cambridge: Harvard University Press, 1965 and later editions), Part 4, and Aaron S. Klieman, *Foundations of British Policy in the Arab World: The Cairo Conference of 1921* (Baltimore: Johns Hopkins Press, 1971).

64 *Words like "Paddy," "Muck-Savage," or "Bog-Wog"* Coogan, 420.

64 *all of them acting together* In the spring of 1998, Sir John Stevens, the head of London's police force, published the results of an official inquiry

that, despite obstruction from Britain's armed forces and Ulster police, had investigated but was unable to prove collusion. See John Mullin, "RUC 'Led Loyalists to IRA Suspects,'" *The Guardian,* March 6, 1999. Subsequently, *The Independent* and *The Guardian* have both published (January 22, 2007) an internal report showing that the secretive Special Branch connived with men in the outlawed Ulster Volunteer Force to murder Irish Catholics and even to deal in drugs and extortion. The charge was admitted by the chief officer of the Protestant police force.

65   *The Church turned against them*   Coogan, 24 and passim.

65   *their struggle against Britain*   This was not then the policy of the Zionist nationalists in the Palestine Mandate. With the approval of the Jewish Agency, terrorist attacks against the British continued throughout the Second World War. There were hundreds of these attacks, but the most spectacular were the attempt on the life of the high commissioner on August 8, 1944, and the murder of British minister of state Lord Moyne on November 6, 1944, for which Winston Churchill bitterly denounced the Zionist authorities in Parliament.

65   *"it is our job to defend them"*   M. L. R. Smith, 85, quoted from S. Cronin, *Irish Nationalism.*

66   *had only ten guns*   Coogan, 279.

66   *Continuity IRA*   Very little is known of this group. As Henry McDonald wrote, "Three held as 'gangland arsenal' is netted in Cork," April 22, 2007, *The Observer,* The "Gardai [The Irish 'Special Branch' police] have smashed a major arms smuggling network involving a Limerick crime gang that cooperates with the Continuity IRA."

66   *nearly four bombings a day*   M. L. R. Smith, 100; some of the information that follows is drawn from the next few pages of this study.

69   *support from the world community*   Coogan, 420.

69   *their "Green Book"*   The text is given in http://www.residentgroups.fsnet. co.uk/greenbook.htm.

70   *began a hunger strike*   A similar but less well-known hunger strike was mounted by prisoners at Guantánamo over their comparable treatment as "unlawful enemy combatants."

70   *was Bobby Sands*   Coogan, 367–368, 377. "The I.R.A. even contacted a Palestinian who had survived a hunger and thirst strike for a lengthy period, so that his experiences could be put to use."

70   *moving episodes in their history*   Coogan, 377. Nine other IRA leaders also starved to death with him.

70   *the leadership of "Óglaigh n hÉireann" announced*   *The Guardian,* July 28, 2005.

71   *arrested by police in actions in Spain, France. and Ireland*   *The Guardian,* June 20, 2006.

71   *untold misery for their followers*   Eamon Quinn and Alan Cowell, "Historic Accord Reached in Northern Ireland," March 27, 2007, *International*

Herald Tribune; Editorial "Days of Decision in Ulster," *Interntational Herald Tribune;* and David McKittrick, "The Hands of History: Two Worlds Come Together to Broker a New Era of Hope," March 27, 2007, *The Independent.*

73 *As one historian has written* Robert Lee Wolff, *The Balkans in Our Time* (Cambridge: Harvard University Press, 1956), 122–123. Mr. Wolff was head of the Balkan Research and Analysis section of the U.S. Office of Strategic Services (the predecessor to the CIA) during the Second World War before becoming professor of history at Harvard University.

74 *the Ustase (the "rebels")* Strongly influenced by fascism, the Ustase had operated military training camps in Italy during the 1930s, and upon returning to Yugoslavia, their militants became the strong-arm of the Croat puppet regime.

74 *in his attempt to invade Greece* Because he desperately wanted the glory of a military victory, because he regarded the Mediterranean as *mare nostra,* the Italian sea, and because he feared that Hitler would prevent his action, Mussolini launched his attack on Greece on October 28, 1940, without informing the Germans.

74 *on the way toward Greece* Hitler had a more serious reason to be furious. The diversion of his armies into the Balkans, while perfect examples of blitzkrieg, cost nearly two months, so when on June 22 German troops invaded Russia, they were unable to reach Moscow before winter set in. Yugoslavia thus set in motion events that were to shape the German defeat.

75 *to German settlers* Robert Lee Wolff, 204.

77 *twenty-six like-minded former officers and soldiers* Fitzroy MacLean, *Tito* (originally published in England as *The Heretic*) (New York: Ballantine Books, 1957), 104. MacLean had access to the Partisans and to British intelligence information and policy discussions on Mihailovic. He is the best non-Yugoslav source on Tito's movement.

77 *the holocaust unleashed upon them by the Germans and Italians* In November 1941 a British liaison officer visited his encampment and found "no remains of Mihailovic's men except himself and a few officers." British Intelligence document, AFHQ, 17 quoted in MacLean, 117.

77 *collaboration with the Germans and Italians* This was the report given to the British government by its emissary to Mihailovic, Colonel William Hudson: that Mihailovic would take no action "before the Axis was on the point of collapse," his forces were "auxiliaries of the Italians to the extent of receiving food supplies and cautious deliveries of arms and ammunition" from them, and in one episode in Dalmatia were used by the occupying powers to kill some seven hundred and wound one thousand Partisans. Colonel Hudson's report is summarized in F. W. D. Deakin, *The Embattled Mountain* (London: Oxford University Press, 1971), 151ff.

78 *in return for supplies* MacLean, 116–117. In his autobiography, *Escape*

to Adventure (Boston: Little Brown, 1950), MacLean recounts a conversation with an Italian colonel who attended a party given at the Cetnik headquarters when one of Mihailovic's senior commanders was awarded a decoration by the Yugoslav king for gallantry in fighting the very Italians who were "carousing at enemy Headquarters." The colonel said "he had always found the Cetnik leader very civilized and easy to deal with." MacLean, *Escape to Adventure*, 267. The original title in England was *Eastern Approaches* (London: Cape, 1951).

78   *the whole world knows as Tito*   Practically nothing was known of Tito during the first part of the war. From British intelligence reports, Churchill thought "Tito" was a Croat terrorist group. And, as Robert Wolff amusingly recounts, some of his colleagues thought that he was a committee known by the acronym "TITO" for Taina (secret) International Terrorist Organization. How they got the acronym from Serbian into English Wolff does not try to explain. Robert Lee Wolff, 210.

78   *later wrote*   MacLean, *Tito*, 93–94. All future references to MacLean refer to this book.

78   *MacLean again*   MacLean, 98.

79   *for his bosses in Berlin*   Quoted by MacLean, 99–100.

79   *As MacLean wrote*   MacLean, 121.

81   *the effect of those events on Tito*   Deakin, 18. Deakin had read modern history at Oxford; then he worked with Winston Churchill helping research one of Churchill's histories; when the war broke out, he joined the army and was seconded to the "Special Operations Executive." He specialized in Balkan affairs and was parachuted into Yugoslavia as the first British officer assigned to the Partisans. There he was in daily contact with Tito and the senior staff of the Partisans. Wounded in a German attack, he shared with the Partisans all the rigors of the war. After the war, he became master of St. Antony's College, Oxford, where I first got to know him. A scholarly man of action, he wrote several books on the Balkans and on the Axis.

82   *"commanded the bulk of our funds"*   MacLean, 71.

85   *"the columns of the hospitals"*   Deakin, 95–96.

86   *Deakin and other British officers*   Several of whom later published their memoirs. One of them, Basil Davidson, *Partisan Picture* (Bedford, England: Bedford Books, 1946), wrote that "The facts of the resistance were splendid enough, and those of the last few months miraculous even . . . [since] In that August [1943] of my arrival there were over 30 enemy divisions on the territory of Jugoslavia . . . My own first impression was one of contrast, a contrast between the utter discrepancy of means available and the end to be achieved." Basil Davidson, 14–15.

85   *"into their component parts"*   MacLean, 109.

85   *sabotage facilities*   As I discussed in the first chapter, partisan or auxiliary war, known in the eighteenth century as *la petite guerre,* is very different

from the politico-military *guerrilla* warfare of insurgency. They have often been confused. The Russian form of partisan war is described in two books by Otto Heilbrunn, *Partisan Warfare* (New York: Praeger, 1962) and *Warfare in the Enemy's Rear* (New York: Praeger, 1963). A personal account is given by Roger Hilsman, *American Guerrilla: My War Behind Japanese Lines* (McLean, Va.: Brassey's, 1990).

86 *regular Turkish troops could not follow* As Colonel T. E. Lawrence ("Lawrence of Arabia") wrote in the 1929 edition of *The Encyclopædia Britannica*, "Camel raiding-parties, self contained like ships, could cruise securely along the enemy's land-frontier, just out of sight of his posts along the edge of cultivation, and tap or raid into his lines where it seemed fittest or easiest or most profitable, with a sure retreat always behind them into an element which the Turks could not enter."

86 *Mosa Pijade, wrote, it was a "fable"* Mosa Pijade, *La Fable de L'aide Sovietique à L'Insurrection nationale Yougoslave* (Paris: Le Livre Yougoslave, 1950).

86 *"in appreciable quantities"* Davidson, 15.

86 *"economic use of what armament you have"* Quoted in Pijade.

87 *into "Red Star" cigarettes* MacLean, 110.

87 *a gift hard for Mihailovic to accept* The guerrillas thus echoed the most ancient form of propaganda: each of the city-states of Greece minted coins decorated with its symbols so that anyone using them in effect confirmed the legitimacy of the city.

87 *its red star prominently displayed* Deakin, 238.

88 *"it gave the character of a revolution"* MacLean, 102.

88 *all the written records, commented* Robert Lee Wolff, 212.

88 *into sending a mission to Tito* Deakin, 255, 263.

90 *Hitler was then doing in Germany* This was not a sudden change. Metaxas had been earlier exiled for his promotion of fascism and was the leader of an ultra-royalist party known with unintended irony as "Free Spirits" (Eleftherophrones). He had been educated at the German military academy in Berlin.

90 *on the Nazi model* Henry Cliadakis, "The Political and Diplomatic Background to the Metaxas Dictatorship, 1935–1936," *Journal of Contemporary History* 14 (1979).

90 *working for the same cause* Hemerologio, IV, 553, cited in John Louis Hondros, *Occupation and Resistance: The Greek Agony 1941–44* (New York: Pella, 1983), 29.

90 *prevented from serving* Stefanos Sarafis, *ELAS: Greek Resistance Army* (London: Merlin Press, 1980 translation of 1946 Greek edition), 21.

91 *for much of its fuel* Hondros, 38–40.

91 *a similar offer to Hitler's ambassador* These events are summarized in Hondros, 48–53.

93 *kill their old political enemies* Led by Colonel George Grivas, the X was

thought to have about five thousand followers; the British later used them against the left-wing guerrillas, supported them with arms and money, but had to fight them eventually in Cyprus.

93   *"taken from the Greeks even their shoelaces"*   Galeazzo Ciano, *The Ciano Diaries, 1939–1943* (New York: Doubleday, 1946), 387 quoted in Hondros, 70.

93   *shocked by what had happened*   C. M. Woodhouse, *The Struggle for Greece 1941–1949* (London: Hurst, 2002 & 1976 edition), 23.

94   *who were still at liberty*   When the war broke out, the KKE was in disarray. Its leader and most of its senior people were in prison, and it is believed that the remaining cadres had been penetrated by Metaxas's political police. Hondros, 109. But during the German occupation, and by the time of the British invasion, the KKE was said to have grown to 350,000. Dominique Eudes, *The Kapetanios: Partisans and Civil War in Greece, 1943–1949* (New York: Monthly Review Press, 1972), 66.

94   *headed by a "responsible person"*   I draw here on L. S. Stavrianos, "The Greek National Liberation Front (EAM): A Study in Resistance Organization and Administration," *Journal of Modern History* 24, no. 1 (March 1952): 42ff.

95   *one in four leaders of EAM was a Communist*   Hondros, 121.

96   *a larger force of "reservists"*   Hondros, 144.

97   *Yugoslav Cetniks under Drazha Mihailovic*   See Chapter Five.

97   *collaborated with the German occupiers*   They have been described by the British officer who was sent to them, Colonel C. M. Woodhouse, who, reflecting British policy, favored EDES and its leader.

97   *denounce him to the Gestapo if he did not*   Woodhouse, 29.

97   *that it be disbanded*   Hondros, 144.

97   *about seven thousand active guerrillas*   Hondros, 144.

98   *never controlled any significant territory*   Gordon Wright, in "Reflections on the French Resistance (1940–1944)," *The Political Science Quarterly* (February 1940): 336ff., accepts the French government figure of 2 percent of adult Frenchmen, but this includes those who joined after Allied landings showed that the Germans had lost the war. As Robert Paxton also points out in *Vichy France: Old Guard and New Order* (New York: Columbia University Press, 1972), it was actually the Vichy government rather than the Germans "who wanted collaboration; Hitler wanted only booty." Paxton, xii. He charges that none of the French resistance groups supported the Free French movement under de Gaulle in the aftermath of the German invasion in 1940 (Paxton, 40) and that "the number of active *Résistants* was never very great, even at the climactic movement of the Liberation" (Paxton, 294–295).

98   *Churchill's strong royalist predisposition*   This hostility is evident throughout the account of the senior British liaison officer, Brigadier E. C. W. Myers, who took a personal dislike to the ELAS commander and regarded EAM as a purely Communist organization, unreliable and unwilling to

follow British orders. *Greek Entanglement* (London: Rupert Hart-Davis, 1955), 101, 114, and passim. Churchill was so reluctant to give the king the decision of the British government that he must issue a statement that he would not return until the Greek people had expressed their will, that he refused to meet him until Anthony Eden came to Cairo. Elisabeth Barker, "Greece in the Framework of Anglo-Soviet Relations, 1941–1947," in Marion Sarafis (ed.), *Greece from Resistance to Civil War* (Nottingham: Spokesman, 1980), 20.

99   *a particularly close relationship with Queen Frederika*   To the point that the general's wife was said to have complained. The queen was Smuts's guest in South Africa while her husband remained in London. John O. Iatrides (ed.), *Ambassador MacVeagh Reports: Greece, 1933–1947* (Princeton, N.J.: Princeton University Press, 1980), 383.

99   *"certainly in Greece and the Balkans"*   Marion Sarafis, 18–19.

99   *established a provisional government*   Richard Clogg, *A Short History of Modern Greece* (Cambridge: Cambridge University Press, 1979), 147.

99   *"miserable Greek banditti" and "treacherous, filthy beasts"*   Churchill sent a personal message to Soviet Foreign Minister Molotov, saying "This is really no time for ideological warfare. I am determined to put down the mutiny . . . I am sure you would not allow such things to go on in the Soviet armies or among any forces which you might control. I therefore hope that Tass may be told to leave off this agitation . . ." Marion Sarafis, 22.

100   *and then on southern France*   *The Second World War: Volume Six: Triumph and Tragedy* (London: Library of Imperial History and Cassell, 1976).

100   *Britain taking charge of Greece*   The Russians agreed, but wanted the deal countersigned by the United States. American Secretary of State Cordell Hull refused, but at Churchill's insistence, Roosevelt (without telling Hull) agreed on June 12. Hull found out about the deal through a leak to the American military attaché in Cairo!

100   *one of the most sweeping deals known in history*   Thus, a year *before* the Yalta Conference (February 7–12, 1945) the great cold warrior himself had been laboring to plant the foundation of what he was later to excoriate as "the iron curtain." Stephen G. Xydis, "The Secret Anglo-Soviet Agreement on the Balkans of October 9, 1944," *Journal of Central European Affairs*, xv (1955) 248 ff. Also see his *Greece and the Great Powers, 1944–1947: Prelude to the Truman Doctrine* (Chicago: Argonaut, 1963).

100   *been hinted at before*   Colonel Deakin, as he then was, met with Churchill, Eden, Field Marshal Smuts, members of the British general staff, and Deakin's opposite numbers in liaison with the Albanians and Greeks when they were passing through Cairo after the Tehran Conference on December 9, 1943. He wrote that "perhaps the Allied landing in the Balkans, which Tito was convinced was about to take place, and at which we had hinted, without being fully aware of the impact of our words, was in some way co-ordinated with Soviet strategic plans and re-

lated to an ultimate partition of South-Eastern Europe." Deakin, 67.

100 *Cordell Hull was strongly opposed* Secretary of State Cordell Hull feared that it was likely to "sow the seeds of future conflict," which, of course, it did. *Memoirs* (New York: Macmillan, 1948), II/1452.

100 *he was prepared to "write off" Yugoslavia* MacLean, *Escape to Adventure*, 309–310.

101 *a remarkable document* *The Second World War: Triumph and Tragedy*, VI/149–150. Churchill recounts that he offered to destroy the handwritten note with Stalin's "tick," but Stalin urged him to keep it.

102 *"establishment of a Communist society"* Quoted in Lawrence S. Wittner, *American Intervention in Greece, 1943–1939* (New York: Columbia University Press, 1982), 10.

102 *"provided the ground is well chosen"* *The Second World War: Triumph and Tragedy*, VI/187.

102 *"let us put an end to the Greek uprising"* Djilas recounts his meeting with Stalin in *Conversations with Stalin* (New York: Harcourt, Brace & World, 1962), but since Djilas did not know of Stalin's deal with Churchill, he commented, "Not even today am I clear on the motives that caused Stalin to be against the uprising in Greece."

102 *Churchill later confirmed* *The Second World War: Triumph and Tragedy*, VI/293.

102 *"where a local rebellion is in progress"* *The Second World War: Triumph and Tragedy*, VI/187. Italics in the original.

103 *killing the disarmed ELAS veterans* Members of the X were later to help to overthrow the Greek government and to establish the dictatorship of "the Colonels" in Greece. On them see Nigel Clive, *A Greek Experience, 1943–1948* (Wilton, England: Michael Russell, 1985), 168ff. Clive was a British intelligence officer in Greece. Their role in overthrowing the later Greek government is the subject of the remarkable film *Z* by Costa-Gavras.

105 *"How We Won in Greece"* *Balkan Studies* (1967): 387ff. He said that "an enlarged, better-equipped, better-trained and better-led Greek Army took over in 1948 and 49, and saved Greece once again." The army overthrew the Greek government in the same year Van Fleet published his article, 1967, and began a rule striking similar to Metaxas's, a brutal, ugly dictatorship, known as the rule of the Colonels.

105 *proved fatal* American officials in Vietnam continued to focus on external aid to the Viet Minh and even today, as I shall point out in discussing Somalia, the senior U.S. official, Assistant Secretary of State Jendayi Frazer, asserts against all historical evidence that "no insurgency group can survive without support from neighboring countries."

106 *engagement of ground troops that was crucial* Like the Viet Minh, the ELAS lived on very little. The amount including food, shoes, blankets, and ammunition a guerrilla consumed was estimated at only about five pounds a day.

108   *the aggressive, nomadic Masai*   The Masai speak a language in a different
        family from Bantu, the Ma dialect of "Nilo-Saharan," and were nomadic
        warriors who lived entirely on their herds. Apparently their only form of
        contact with the Bantu-speaking peoples of Kenya was cattle raiding; they
        did not engage in trade or attempt to seize land. The other, smaller, tribes
        played lesser roles.

108   *Eastern European Jews*   On August 14, 1903, at the urging of Joseph
        Chamberlain, the British government, in a letter remarkably like the later
        Balfour Declaration, offered the Zionist organization six thousand square
        miles of Kenya for settlement with "a free hand in regard to municipal
        legislation as to the management of religious and purely domestic matters,
        such local autonomy being conditional upon the right of his Majesty's
        Government to exercise general control." Settlement in Kenya would not
        have replaced the movement toward Palestine, as Max Nordeau then ar-
        gued, but it would create a *Nachtasyl* (overnight refuge), that would take
        the pressure off the Jewish community in Russia. Theodor Herzl favored
        the scheme, but he was unable to overcome the resistance of those who
        wanted to focus all attention on Palestine. On the Jewish side, it met with
        fatal opposition from the Sixth Zionist Congress, and on the English side,
        the newly arrived white settlers in Kenya were furious. On this fascinat-
        ing "might have been" bit of Kenyan and Middle Eastern history, which
        would have pitted the later black resistance movement against the Zion-
        ists, the essential story is told by an early participant, Richard Gottheil,
        in *Zionism* (Philadelphia: Jewish Publication Society of America, 1914),
        123ff. Also see M. P. K. Sorrenson, *Origins of European Settlement in Ke-
        nya* (Nairobi: Oxford University Press, 1968), 36.

110   *enforce decisions on their tribesmen*   Between the whites and the blacks
        was an Indian community, in which the Ismailis were prominent, who
        were mainly shopkeepers and merchants. Descendants of South Asians
        who had been imported to build the railways and those who supported
        the workers, they then numbered about one hundred thousand. The In-
        dians wisely kept their heads down but received a sort of representation
        at least to the government in London by the leader of the Ismaili com-
        munity, the Aga Khan. They were largely irrelevant to the Kikuyu-settler
        struggle.

111   *to become its president*   David Anderson, *Histories of the Hanged: The
        Dirty War in Kenya and the End of Empire* (New York: Norton), 2. An-
        derson is a noted authority on Africa and lecturer in African studies at
        Oxford University. I have relied heavily upon him and Caroline Elkins,
        author of *Imperial Reckoning: The Untold Story of Britain's Gulag in Ke-
        nya* (New York: Henry Holt, 2005) because of the access they had to
        official documents on both the insurgency and the British counterinsur-
        gency. Their excellent and detailed books make previous accounts of the
        insurgency no longer acceptable. Kenyatta tells his own story in *Facing*

111   *Mount Kenya* (London: Secker & Warburg, 1938 and later editions).
        David Anderson wrote   Anderson, 1.

113   *Kariuki recounts*   *Mau Mau Detainee* (Oxford: Oxford University Press, 1963), 52ff.

113   *mocked with the name of "Mau Mau"*   The words "Mau Mau" were said to have been made up by Kikuyu, who "often make puns and anagrams with common words," but Jomo Kenyatta repeatedly said that they were unknown among the Kikuyu.

114   *"horror of Mau Mau oathing"*   Anderson, 50–51.

115   *The newly arrived governor, Sir Evelyn Baring*   Baring might be said to have been born for an imperial role in Africa since he was the son of the master of Egypt under the British Empire, Lord Cromer. He had served in the Indian Civil Service and later became the high commissioner of South Africa.

115   *"a terrible hopeless position"*   Aikins, 33ff.

116   *less-well-known younger men*   Kenyatta was immediately flown off to a remote detention camp where he would spend most of the insurgency before being brought back and used by the British to end it.

116   *in other insurgencies*   A typical example is H. P. Willmott, "Kenya in Revolt: The Mau Mau Troubles Involved Military Operations Throughout Central Kenya," in Robert Thompson (ed.), *War in Peace: Conventional and Guerrilla Warfare Since 1945* (London: Orbis, 1981), 108ff.

116   *"the grassroots depth of the movement"*   Elkins, 37–38.

116   *"five shillings a nob"*   Described in a British document, Public Record Office, CO 822/489/20, by Frederick Crawford on March 16, 1953, noted in Elkins, 51.

116   *the end of British rule*   Elkins, 51.

117   *"used against his own people"*   Anthony Clayton, *Counter-Insurgency in Kenya: A Study of Military Operations against Mau Mau* (Nairobi: Transafrica Publishers, 1976), 38–39, quoted in Elkins, 52.

117   *"even six months ago"*   Public Record Office: CO822/441, appendix X quoted in Wunyabari O. Maloba, *Mau Mau and Kenya: An Analysis of a Peasant Revolt* (Nairobi: East African Educational Publishers, 1994), 83.

117   *develop a new plan*   I rely for much of the following upon Elkins, 100ff. The quote is from her book.

118   *Robert (later Sir Robert) Thompson*   Sir Robert Thompson's works on Malaya and Vietnam include his *Defeating Communist Insurgency* (New York: Praeger, 1966). Thompson's book was produced under a CIA contract to assist Americans in applying the Malayan model to Vietnam.

119   *structure of his guerrilla force*   The interrogation is cited from the Johnson Papers, mss Afr.S. by Maloba, 84–85.

119   *swept from their homes*   Maloba, 86. Subsequent pages substantiate the following information on the counterinsurgency.

119   *in concentration camps*   Since the prisoners were neither charged nor tried, this was in violation of the European Convention on Human Rights to which Britain was a party. Consequently, it caused a *crise de conscience* in England then rather as Guantánamo has in America in more recent times.

121   *forced on the reluctant governor*   It led to the "million-acre program" for the settlement of about seventy thousand families. Maloba, 162ff.

122   *wean them away from the movement*   Louis Leakey, *Defeating Mau Mau* (London: Methuen, 1954).

122   *White African*   Louis Leakey, *White African* (London: Hodder & Stoughton, 1937).

122   *"a blunt, brutal and unsophisticated instrument of oppression"*   Anderson, 7.

125   *Algiers by the then ruler*   A similar motivation had driven the British invasion of Bengal. P. J. Marshall, *Bengal: The British Bridgehead, The New Cambridge History of India* (Cambridge: Cambridge University Press, 1987) II/2, 77, 83–92, and Stanley Wolpert, *A New History of India* (Oxford: Oxford University Press, 1989), 187–189. Both invasions were among the few in history that were financially profitable. From Algeria, the French government took what in today's terms amounted to more than $1 billion in gold and jewels and destroyed perhaps as much as they took. The episode is recounted with verve in Wilfrid Blunt, *Desert Hawk: Abd el Kader & the French Conquest of Algeria* (London: Methuen, 1947), 15.

126   *from Spain in 1492*   L. P. Harvey, *Islamic Spain: 1250–1500* (Chicago: University of Chicago Press, 1990), 321, 324–325.

127   *"revolt against the central authority"*   William Spencer, *Algiers in the Age of the Corsairs* (Norman: University of Oklahoma Press, 1976), 57. The information above on the population is drawn from this excellent book.

127   *"gay, outlandish uniforms"*   Blunt, 11.

128   *"it had virtually subjugated"*   Blunt, 27.

128   *Arabic did not then have a word*   I discuss in my book *The United States and the Arab World*, chap. 8, the origin and elaboration of the concept of nationalism. The early word for nationalism derived from the word *watan* which, like the French *pays,* then meant just "village" or "district." Among the nomads, who were not so rooted to a locality, the concept was based on the notion of the kinship group, the clan (*qawm*), which lived, herded, and fought as a unit. Neither group thought of itself as part of a larger whole, a nation-state.

128   *every step they took for fifteen years*   Although dated, perhaps the best biography of him in English remains Wilfrid Blunt's spirited book.

128   *Charles Collwell wrote   Small Wars: The Principles and Practices* (1st ed., 1896, 3rd [this ed.], London: HMSO, 1906, reprinted Lincoln: University of Nebraska Press, 1996, with a new introduction by Douglas Porch, professor of strategy at the Naval War College, Newport, R.I.), 128. As Collwell wrote (21), he did not mean by "small wars" wars that were small in scale but colonial wars that pitted European armies against natives

rather than the formal wars of mass that took place typically in Europe.

129 *"cumbrous column could not follow it"* T. E. Lawrence described the Arab tactics in the revolt against the Ottoman Empire in his article on guerrilla warfare in the 1925 *Encyclopædia Britannica.*

129 *"for at least a fortnight"* Blunt, 169–170, 171.

130 *he surrendered* Blunt, 216ff.

131 *Alexis de Tocqueville* Known to Americans mainly for his brilliant portrait of American society, *Democracy in America,* who after his tour in America had been elected to the French National Assembly.

131 *"became acquainted with us"* Quoted in Alistair Horne's excellent account of Algeria, *A Savage War of Peace: Algeria 1954–1962* (London: Macmillan, 1977), 29. As head of the American interdepartmental task force on Algeria during the last part of the war, I had access to all the American intelligence reports (most of which were provided by the French), but I have found Horne's account both more consistent and more insightful, so I will often cite it in this account. Another excellent study that I have often consulted here on the latter part of the insurgency is William B. Quandt, *Revolution and Political Leadership: Algeria, 1954–1968* (Cambridge: MIT Press, 1968). It is particularly valuable because Quandt interviewed many of the key figures.

131 *on behalf of the American government* As the member of the Policy Planning Council responsible for the Middle East and North Africa, I was head of the interdepartmental task force on Algeria.

132 *"soon create one for themselves"* Quoted in Horne, 37.

132 *influenced by similar revivalist movements* The movements called themselves *salafiyah.* Like the European Puritans, the *salafis* sought to lead their peoples forward by reasserting the fundamental principles of their religion. The word thus has various and, to English speakers, contrary meanings: it connotes both the "the vanguard," that is, those who lead, and also "traditionalists," those who preserve the inheritance of the past.

133 *an entity as the Algerian nation* As he said, "Had I discovered the Algerian nation, I would be a nationalist . . . However, I will not die for the Algerian nation, because it does not exist . . . One cannot build on the wind." Quoted in Horne, 40.

133 *marginalized as the "token Algerian"* On Abbas and his group, see Quandt, chap. 3 and 52ff.

133 *married a French woman* His wife was or became a member of the Communist Party in France.

134 *and the party was banned* In 1945, Messali's sentence was reduced to house arrest, and the party reemerged under a different name.

134 *"in a state close to peonage"* Manfred Halpern, "The Algerian Uprising of 1945," *The Middle East Journal* 2 (1948): 193.

134 *and the French army* Quandt, 51.

134 *to fight the Germans* French use of American-supplied war matériel

against the insurgents in both Vietnam and Algeria would be a constant source of friction between the French and American governments and would affect American relations with others for decades.

135   *in front of a firing squad*   Statement by the leader of the Algerian Communist Party, Amar Ouzegan, in the party newspaper *Liberté,* quoted in Horne, 27.

135   *were above all else Europeans*   Halpern, 192.

135   *French soldiers to fight the Algerians*   Horne, 136–137.

135   *fled abroad*   They are named and discussed in Horne, 75–79.

136   *defeated by a colonial people*   I will discuss it in the next chapter.

136   *on the colonial Americans*   See above Chapter One. Bigelow, 189–190.

136   *General Jacques Massu*   Perhaps the best portrayal of this emotional transformation is given in two novels by Jean Lartéguy, *The Centurions* (New York: Dutton, 1962) and *The Praetorians* (New York: Dutton, 1963). For a more factual account, see Martin Windrow, *The Last Valley: Dien Bien Phu and the French Defeat in Vietnam* (New York: Da Capo, 2004), 655–657, and Geoffrey Bocca, *The Secret Army* (Englewood Cliffs, N.J.: Prentice-Hall, 1968), Part I.

136   *John Paul Vann*   On whom see below, Chapter Ten, Vietnam.

137   *the war of independence had begun*   Having merged Algeria into France, the French had discouraged the use of Arabic even for street signs. Arabic did not become an official language until 1947. The psychology of denationalization is enormously complex. Attempting to unravel it was the task Frantz Fanon set for himself in *Les Damnés de la Terre* (Paris: Maspero, 1961, translated as *The Wretched of the Earth,* New York: Grove, 1965) and other works.

137   *the few dozen in 1954*   Numbers are impossible to verify and vary widely. Rabah Bitat, a member of the first *comité de direction* of the movement claims nine hundred "ready to act" as of November 1, 1954. Quandt, 93.

137   *to get international recognition*   Horne, chap. 5.

138   *"destroys, bombards, summarily executes"*   Commandant Vincent Monteil in a letter to Jacques Soustelle in June 1955. Horne, 117.

138   *two Kabyles*   The Kabyle people are the Algerian Berbers, speakers of the Zwâwah dialect of the Berber language who inhabit the Kabylie mountains in eastern Algeria.

138   *the militants*   With unintended irony, the Algerian nationalists adopted the French term *militants* for themselves.

139   *"consider yourselves as traitors!"*   Quoted in Horne, 142.

139   *only of the "interior" leaders*   Ben Bella along with four of his close "external" associates was arrested by the French on October 22, when his plane, piloted by a Frenchman, was forced down in Algeria; they were imprisoned in France for the next five years. Thus, effectively, direction of the resistance was in the hands of the "internal" leaders and particularly Ramdane Abane throughout most of the rest of the war. Ultimately he was to be killed by his FLN comrades.

139	*divided into six regions* Quandt, 92.

139   *divided into six regions*   Quandt, 92.

140   *the little Bosnian town of Jajce*   The British liaison officer F. W. D. Deakin was there to watch. As he wrote, "We were the silent spectators of the birth of a regime." Deakin, 257. No such outside observers were present at the shepherd's hut in the Kabylie.

140   *the Greek EAM regarded Zervas*   Quandt, 97.

140   *the Algerian war for independence*   There were many incidents, some very bitter, between these great events. In June 1956 the French guillotined two leaders of the FLN despite the fact that one was crippled. Because they were popular and had demanded prisoner of war status, the Algerians saw their deaths as murder. Ramdane Abane ordered reprisals. I do not think, however, that this event influenced his decision on the battle of Algiers; that would have taken place, I believe, even if Governor-General Robert Lacoste had granted clemency.

141   *to apply torture*   The book justifying torture by the French paratroop officer Roger Trinquier, *La Guerre Moderne,* was published in English (New York: Praeger, 1962) under a CIA contract for use in Vietnam.

141   *kill one after another the operational chiefs*   The story of the battle is so meticulously recounted in Gillo Pontecorvo's film *La Battaglia di Algeri* (*The Battle of Algiers*) that it has been used as a training course for American troops operating in Iraq.

141   *The hideous record of torture*   One of the popular French books of the period called it "the cancer of the nation."

141   *almost cost de Gaulle his life*   Two attempts are known. The second and more serious was organized by French Air Force Colonel Jean-Marie Bastien-Thiry with members of the Organisation de l'Armée Secrète (OAS). Finally, when the OAS began shooting French army soldiers, the army turned on some of its most decorated officers who were members of the OAS. See Horne, 524.

141   *literally blowing up Algeria*   The most satisfactory published account of the OAS is Bocca.

142   *who were aiding the OAS*   The OAS not only continued to attack the Muslims but also began to attack the French army and gendarmes. Furious, the French commander sent twenty thousand soldiers equipped with artillery and tanks and aided by aircraft against the European quarter of Algiers where the OAS fighters were hiding out. Embarrassingly, this clash happened in March, while the Evian negotiations were in progress. Casualties were surprisingly light, but the whole district of Bab el-Oueld was virtually destroyed. A confused period of triangular fighting ensued. Finally, in the last week of the war, when I was in Algiers, American intelligence discovered that information on about sixteen thousand OAS *colons* was revealed by the French to the Algerians, who liquidated them. De Gaulle apparently was determined that they not go to France. Some of these events are covered in Horne, 512ff.

144   *more studied than Vietnam*   Vietnam is properly two words: Viet is the name of the people and Nam means south or southern. The Viet people also existed in China; those with whom we are concerned here are the people who moved south into the hinterland of the South China Sea. On the lack of an analytical and deep historical treatment of Vietnam even after so many years of foreign involvement, see Joseph Buttinger, *A Dragon Defiant: A Short History of Vietnam* (New York: Praeger, 1972), preface.

145   *"as an Indian is of caste"*   Douglas Pike, *Vietcong: The Organization and Techniques of the National Liberation Front of South Vietnam* (Cambridge: MIT Press, 1966), 6. I have depended on Pike; John T. McAlister, Jr., *Vietnam: The Origins of Revolution* (New York: Knopf, 1969), chap. 2; Joseph Buttinger, *A Dragon Embattled*; and Paul Mus, "The Role of the Village in Vietnamese Politics," *Pacific Affairs* 28 (1949): 265ff., for much of the following information on the early history and social structure of Vietnam.

146   *millions of aliens over its long history*   For a quick synopsis see my *Neighbors and Strangers: The Fundamentals of Foreign Affairs* (Chicago: University of Chicago Press, 1997), 314–315.

146   *"for the rest of my life"*   *Pentagon Papers*, I/49.

147   *no further attention to the country*   R. E. Latham (trans.), *The Travels of Marco Polo* (Harmondsworth, Middlesex: Penguin, 1958), 221–223.

147   *the tanks of that period, five hundred*   Much of the following information is based on two works of Joseph Buttinger—a three-volume history, *The Smaller Dragon: A Political History of Vietnam* (New York: Praeger 1958) and a two-volume history, *A Dragon Embattled*—and on D. G. E. Hall, *A History of South-East Asia* (London: Macmillan, 3rd ed., 1968).

148   *the Vietnamese "prince" Ngo Dinh Diem*   See Chapter Three. Lansdale met with Diem "almost nightly." *Pentagon Papers*, I/230.

151   *clandestine organization*   This is an insight particularly emphasized by Pike, 8–11.

152   *"disappear into nowhere"*   Admiral Louis-Adolphe Bonard, quoted in Stanley Karnow, *Vietnam: A History* (New York: Viking-Penguin, 1991), 119.

152   *"Enlist that class in our interests"*   Governor-General Jean-Marie de Lanessan to the later Marshal Louis Lyautey, quoted from André Maurois, *Marshal Lyautey* (London: John Lane Bodley Head, 1931), in Robert B. Asprey, *War in the Shadows: The Guerrilla in History* (New York: Doubleday, 1975), 219.

152   *manufacture guns and ammunition*   Buttinger, *A Dragon Embattled*, I/126 and Karnow, 121.

152   *for the Viet insurgents was "pirate"*   For example, see a French dispatch of March 20, 1946, discussing guerrilla bands of up to one thousand men that were fighting against them, reprinted in *Pentagon Papers*, I/18.

153   *"voluntary collaborators"*   Lyautey and Albert de Ponvourville, quoted in Buttinger, *A Dragon Embattled*, I/84.

153   *"expansion of French control"*   Buttinger, *A Dragon Embattled,* 479; Asprey, 224; *Pentagon Papers,* II/142. The author of "Developing a Consensus among the Advisors" comments that although the term itself was avoided, "the process and rationale he [Assistant Secretary of State Roger Hilsman] put forth were the same . . ."

153   *proclaimed their legitimacy*   Karnow, 121, Buttinger, *A Dragon Embattled,* I/130.

153   *"beheaded all his prisoners as 'rebels'"*   Karnow, 98, Buttinger, *A Dragon Embattled,* I/130.

154   *after one of his oil spots had been secured*   Buttinger, *A Dragon Embattled,* I/137–138.

154   *and arrested its teachers*   Buttinger, *A Dragon Embattled,* I/53–54.

155   *"the path chosen by Japan"*   Buttinger, *A Dragon Embattled,* I/151.

155   *against the colonial administration*   Hall, 758ff.

155   *Pulo Condore (Con Son) island*   Buttinger, *A Dragon Embattled,* I/54, 100, 220.

156   *landless laborers*   Known in Vietnamese as *tadien,* who "had not existed in precolonial Vietnam." Buttinger, *A Dragon Embattled,* I/164.

157   *a "well-accepted" policy*   William Roger Louis, *Imperialism at Bay: The United States and the Decolonization of the British Empire, 1941–1945* (New York: Oxford University Press, 1978), 198.

157   *foreign armies of occupation*   Louis, 259.

157   *nearly half a century later*   Karnow, 158. Perhaps the most valuable feature of Karnow's book is that he records interviews he had with a number of the Vietnamese leaders. Giap proudly told Karnow that the French after-battle report "stated that our troops were brave and disciplined—and that their leader displayed a mastery of guerrilla tactics. *Quel compliment!*" They would not be so generous in the future.

157   *"the sole efficient resistance apparatus within Vietnam"*   *Pentagon Papers,* I/45.

158   *described as "modest" aid*   *Pentagon Papers,* I/7.

158   *how to use them*   Karnow, 149–150.

158   *average contingent of Japanese forces*   McAlister, 110.

158   *the next quarter century*   Since Ho obviously knew what he had written, the only group to be kept in the dark was the public in the Allied countries. The documents are alluded to in *Pentagon Papers,* I/20 and I/50. They are also briefly mentioned in Neil Sheehan, *A Bright Shining Lie: John Paul Vann and America in Vietnam* (New York: Random House, 1988), 152–153. It was Sheehan, then a reporter for the *New York Times,* who brought the *Pentagon Papers* before the American public after they had been secretly photocopied by Daniel Ellsberg. To overcome an injunction obtained by the Justice Department to stop the *Times* from publishing them, Senator Mike Gravel of Alaska read a part of the papers into the Congressional Record, thus releasing them to the

public. Sheehan's book, *A Bright Shining Lie,* was begun while he was a fellow of the Adlai Stevenson Institute while I was its president.

159   *"but no French"*   Karnow, 163. Unknowingly, he was echoing what the Syrians had said to the Americans thirty years before when the French set out to conquer that country.

159   *"the reestabilishment of French control"*   Louis, 552.

159   *"containment of communism"*   It was formally set out somewhat later in National Security Council Memorandum 64 in 1950, which warned that "the neighboring countries of Thailand and Burma could be expected to fall under Communist domination if Indochina were controlled by a Communist-dominated government. The balance of Southeast Asia would then be in grave hazard." *Pentagon Papers,* I/83.

159   *than the French right*   Later they so changed their position on Vietnam, as they also did on Algeria, that French Communist dock workers refused to unload caskets of French soldiers killed in Vietnam. Personal communication from Peter Scholl-Latour.

160   *"the execution of contracts"*   *Pentagon Papers,* I/23.

161   *"or capture supplies"*   McAlister, 218 and *Pentagon Papers,* I/25. France "discovered that its military forces were incapable of controlling even the principal lines of communication . . ."

161   *the only recognized national leader*   *Pentagon Papers,* I/49.

161   *Paul Mus called the "Confucian balance"*   "Viet Nam: A Nation Off Balance," *The Yale Review* 41 (1952).

162   *a member of at least one*   McAlister, 263.

162   *"want a fight, they'll get it"*   Karnow, 146, 211. His interview with Giap is illuminating on the Viet Minh strategy. *Pentagon Papers,* I/21.

162   *"within fifteen months"*   *Pentagon Papers,* I/67.

162   *Privately, he was sure he could not*   Private communication from noted French journalist Peter Scholl-Latour who interviewed him at that time.

162   *graveyard of the French Empire in Indochina*   Remarkably recounted by Bernard B. Fall, *Hell in a Very Small Place* (Philadelphia: Lippincott, 1967) and Windrow. It was captured in the film *Diên Biên Phu* by Pierre Schoendoerffer.

163   *"cutting and running"*   Richard J. Barnet, *Intervention and Revolution* (New York: New American Library, 1968), 182, drawing on a personal communication from Bernard Fall, who found the reference in the U.S. Naval Archives.

163   *The Japanese refused*   President Roosevelt to the Japanese ambassador, *Pentagon Papers,* I/8–9. I will rely heavily on the *Pentagon Papers* in this chapter. The papers were assembled by a team in the Department of Defense who were given access to the most confidential records then extant in the American government. Not only are they voluminous and detailed but, since they were not intended for use outside of a small group of senior officials authorized to see every category of restricted materials,

they are more frank and objective than public documents or declarations. They also contain classified papers not otherwise available. Although they show examples of haste and are surprisingly poorly edited, they are the best insights into what was then known that we have. They exhibit, of course, the weakness that they cannot show more than officials then believed. Some of what they knew or believed has turned out to be wrong. And they cannot reveal the frequent verbal communications that form the basis of so much "inside" decision making. These weaknesses are true of all diplomatic archives and inevitably also of all diplomatic history. Consequently, where possible and useful, I have compared them to other accounts. They have been criticized, not so much for the materials they contain as for the prejudices and proclivities of the authors, by a number of writers, including J. L. S. Girling, *The* Pentagon Papers: *The Dialectics of Intervention* (London: Royal Institute of International Affairs, 1972), 61ff. and, in vol. 5 of the Beacon Press edition of the papers, articles by Noam Chomsky, Howard Zinn, Gérard Chaliand, Truong Buu Lam, and others. Most of the comments on the papers focus on the intent of the government officials rather than on their understanding of what I am primarily concerned with here, the processes of guerrilla warfare and counterinsurgency.

164   *"flew the French flag in 1939"*   Various communications to the government of Vichy in November 1942, quoted in *Pentagon Papers*, I/2–12.

165   *"welcome amicably the French Army"*   The cease-fire agreement is quoted in *Pentagon Papers*, I/18–19.

165   *"excluded meaningful negotiations"*   *Pentagon Papers*, I/53.

165   *"guerrilla warfare may continue indefinitely"*   *Pentagon Papers*, I/29.

166   *"Communists could capture control"*   *Pentagon Papers*, I/31.

166   *proclaimed under Emperor Bao Dai*   *Pentagon Papers*, I/ 32–41.

166   *"U.S. policy remained unillumined"*   *Pentagon Papers*, I/32–52, 246.

167   *the United States took no action*   *Pentagon Papers*, I/97. The authors comment that "no record of Operation Vulture has been found in the files examined." They further state that "there is no indication that there was any serious consideration given to using nuclear weapons at Dien Bien Phu or elsewhere in Indochina." *Pentagon Papers*, I/100. But for the French attempt to get them to do so, see I/461.

167   *the French military establishment there, $1.063 billion*   *Pentagon Papers*, I/54–55, 77.

167   *America's ability to meet its NATO obligations*   *Pentagon Papers*, I/92–93.

168   *"he has double-crossed us"*   Karnow, 218–220.

168   *the South (with American support) refused*   *Pentagon Papers*, I/241.

168   *civilians favorable to them moved north*   *Pentagon Papers*, I/247.

169   *"to capture or death"*   Sheehan, 374–375.

169   *as Vo Nguyen Giap boasts*   *"Big Victory. Great Task"* (New York: Praeger, 1968), 52.

169 *"Do what you did in the Philippines"* Sheehan, 134.

170 *"the creation of Edward Lansdale"* Sheehan, 141.

170 *used the fascist salute* Sheehan, 179.

170 *arrived in November 1954* Pentagon Papers, I/181–182.

171 *American critics swung to his support* Pentagon Papers, I/182–183.

171 *with alien, mostly Catholic, appointed officials* Pentagon Papers, I/255.

171 *"the resistance against the French"* Pentagon Papers, I/309.

171 *favorable to the Viet Minh* Pentagon Papers, I/295.

171 *"symbol[s] of insecurity and repression"* Pentagon Papers, I/306.

171 *senior officials were soldiers* Pentagon Papers, I/257.

172 *"220,000 acres of wilderness"* Pentagon Papers, I/255.

172 *on which his "personality" depended* Frances FitzGerald, *Fire in the Lake* (Boston: Little Brown, 1972), 143–144.

173 *building their organization* Pentagon Papers, I/258.

173 *assassinating the appointed village officials* It was not until nearly three years later, when it was evident that the Southerners were determined come what may to press ahead, that Ho's government began to play a significant part.

174 *and four thousand from 1960 to 1961* Pentagon Papers, I/336.

174 *"health workers, agricultural officials, etc."* Pentagon Papers, I/334–335.

174 *"with a view to establishing bases"* This quotation is drawn from a captured document, the "Military Plan of the Provincial Party Committee at Baria," Publication 7308 of the U.S. Department of State, December 1961, Part II, The Appendices, 101.

174 *guerrilla warfare in Vietnam* As mentioned in the introduction, I summarized my findings in a lecture to the "best and brightest" at the colonel level of the army, navy, air force, and marine corps at the National War College.

175 *did not expect to have to resort to force"* Pentagon Papers, I/248.

175 *contrary to the instructions from the party* Pentagon Papers, I/260, quoting George McTurnin Kahin and John W. Lewis, *The United States in Vietnam* (New York: Dial, 1969).

175 *"Soon the fighting forces quadrupled to 10,000"* Sheehan, 194, 195.

175 *all the intelligence sources* Pentagon Papers, I/242. The editors comment that they drew upon all intelligence reports "including the most carefully guarded finished intelligence," raw intelligence, and open sources.

175 *"a single capital; a single country"* Sheehan, 196. At this time, W. W. Rostow, deputy director of the National Security Council, argued that Vietnam was "a pretty straight-forward case of external aggression." *Pentagon Papers,* II/98. He was using the slogan "Ho Chi Minh Go Home." Of course, Ho was at home and the Viet Minh movement was acting as a single unit.

176 *"they should pursue in attaining them"* Pentagon Papers, I/262.

176 *"training facilities built and fortifications dug"* "Talking Paper for the Chair-

man of the Joint Chiefs of Staff for a meeting with the President on January 9, 1962," published as Document 105 in *Pentagon Papers*, II/654–655.

177   *the war in 1961 had been Americanized*   Pentagon Papers, I/251ff.

177   *usually intent on avoiding combat*   Sheehan, 683, 742–743 and passim on John Paul Vann's experience in association with army units.

177   *Special Forces troops did not work*   Colonel (later Lieutenant General) Fred Haynes, USMC, who was a member of the task force I headed, went to Vietnam as chief of staff of the First Marine Division. When I asked him about the Special Forces "stiffening," he laughed and commented that whenever the South Vietnamese troops were informed of operations, the marines inevitably ran into an ambush. Political-intelligence penetration, as we later learned, went throughout the army general staff. The marines, Haynes said, fought on their own if they wanted to live.

177   *"win the war with the U.S. Army"*   Sheehan, 556.

177   *under the guidance of Robert Thompson*   He tells of his experience in *Defeating Communist Insurgency*. Few then questioned whether the experience was relevant. In Malaya, the insurgents were a maximum of about ten thousand drawn from the Chinese minority community, which the Malays considered to be foreigners. The government forces had a twenty-to-one advantage in fighting men. Thompson's key idea was to deny the insurgents food and shelter by "regroupment." So extreme was the British program that people in the regroupment centers were forced to eat in community messes to prevent them from having any food that they might pass to the insurgents. The British also expelled from the country thousands of Chinese. They could do so since the insurgents were foreigners; in Vietnam, the government obviously could not expel native peasants. In a word, the British strategy in Malaya was irrelevant.

178   *sixty-eight hundred barbed-wire-encircled strategic hamlets*   Sheehan, 338ff.

178   *"all failed dismally"*   Pentagon Papers, II/128.

178   *"ceased to exist"*   Sheehan, 373.

178   *"revolt against the GVN [government of South Vietnam]"*   Pentagon Papers, I/256.

179   *"fired it off themselves"*   Sheehan, 372–373.

179   *stopped using Soviet equipment*   Sheehan, 67, 69, 100, 372, 374.

179   *along the fifteen-hundred-mile coastline*   Pentagon Papers, IV/110.

179   *highly sophisticated equipment available to the Americans*   Partly described in James W. Gibson, *The Perfect War: Technology in Vietnam* (Boston: Atlantic Monthly Press, 1986), 396ff.

180   *led to their targets by Southern guerrillas*   There, the guerrillas reverted to the eighteenth-century role of informal troops, what had been known as *la petite guerre*, as auxiliaries rather than independent guerrilla units.

181   *cold war against each other*   The subject of an engrossing book by Peter Hopkirk, *The Great Game* (Oxford: Oxford University Press, 1990).

182   *"enough mountains in the Soviet Union already"*   In a letter to *The Times*,

quoted in Nigel Ryan, *A Hitch or Two in Afghanistan: A Journey Behind Russian Lines* (London: Weidenfeld & Nicolson, 1983), 153.

182   *spread Communism to surrounding countries*   This is slightly unfair to John Foster Dulles. Although he believed it, he did not conceive it. The domino theory was first spelled out in America during the Truman Administration before the Korean War. It was incorporated in National Security Council (NSC) memorandum 64 and was adopted as U.S. policy on February 27, 1950. As the editors of the *Pentagon Papers* write (I/83–84) "The 'domino principle' in its purest form was written into the 'General Considerations' section of NSC [National Security Council memorandum] 124/2. It linked the loss of any single state of Southeast Asia to the stability of Europe and the security of the United States."

182   *infecting them with anti-Communist aspirations*   Eqbal Ahmad and Richard Barnet, "Bloody Games," *The New Yorker,* April 11, 1988.

184   *the province of Qandahar*   Popularly but incorrectly transliterated Kandahar.

186   *neglected one element, the Afghan people*   To counter the British assumption of power in Afghanistan, the Russians sent a force of five thousand troops against the khan of Khiva. Only one in every five men made it out alive. Unguessed at the time, the Russian catastrophe at Khiva was an augury of the fate of the British at Kabul.

186   *"increased hatred and distrust"*   Louis Dupree, *Afghanistan* (Princeton, N.J.: Princeton University Press, 173), 378. Dupree was the foremost American authority on Afghanistan and was my guide during a lengthy tour I made of Afghanistan in 1962 when I wrote a U.S. government policy study on the country. We went over the route of march of the Anglo-Indian force and its retreat.

186   *Stanley Wolpert has written*   *A New History of India* (New York: Oxford University Press, 1977), 220.

186   *"perfectly wonderful how they hang together"*   Secret Letters from India, vol. 82, 9 Jan. 1842, No. 9, India Office Records, London quoted in Dupree, 386.

187   *the most impressive guerrilla war of the century*   The British sent a column back to Kabul, took their revenge, and settled down for six months. But then they did not know what to do with their "victory." In despair, the governor-general of India on October 1, 1842, instructed the commander of his troops as follows: "The British armies in possession of Afghanistan will now be withdrawn . . . The Governor-General will leave it to the Afghans themselves to create a government . . . To force a sovereign upon a reluctant people would be as inconsistent with the policy as it is with the principles of the British Government, tending to place the arms and resources of that people at the disposal of the first invader, and to impose the burden of supporting a sovereign, without the prospect of benefit from his alliance . . . The enormous expenditure required for the support of a large force in a false military position at a distance from its

own frontier and its resources, will no longer [be our policy]."

188   *"perceived that we were not invincible"*   George Pottinger, *The Afghan Connection* (Edinburgh: Scottish Academic Press, 1983), 206.

188   *"the greatest difficulty is to know where to stop"*   Quoted in Firuz Kazemzadeh, "Afghanistan: The Imperial Dream," *New York Review of Books,* February 21, 1980.

189   *Afghanistan north of Herat*   Later incorporated into Turkman SSR, which subsequently became Turkmenistan.

190   *in the previous administration*   One was an irrigation project on the Helmand river that had been developed with inadequate studies of the ground, so lack of drainage proved ruinous, while inattention to the needs of the farmers for credit kept them in unproductive poverty; a second project was a superhighway linking Kabul to Qandahar for which there was little need and which was not supported by farm-to-market subsidiary roads; a third project was a large airport at Qandahar for which there was no traffic; and most bizarre of all was a scheme to exploit a coal mine in the center of the vast Hindu Kush mountain range, which required more expenditure of energy to transport its product to where it could be used than the coal would yield. Afghanistan was a blank on the American policy map.

191   *Pukhtunwali, illuminated and guided their lives*   James Spain, *The Way of the Pathans* (London: Trinity Press, 1962), 46ff.

192   *the opinions of the community are sorted out*   A similar institution, known as the *panchayat,* exists in northern India.

192   *no power outside his own group*   Only on rare occasions and for limited purposes, *maliks* might assemble in a regional or Afghan-wide assembly (*loya jirga*).

193   *a tribal rebellion ten years later*   He made three mistakes that the Russians and their local allies would make in the 1980s: he tried to stop the traditional toll (Pushtu: *badraga*) on transit through tribal territories; he pushed various alien innovations; and he alienated the Muslim religious leaders.

193   *"undisciplined and hungry for loot"*   Sir W. Kerr Fraser-Tytler's description, quoted in Vartan Gregorian, *The Emergence of Modern Afghanistan* (Stanford: Stanford University Press, 1969), 285.

194   *Soviet troops*   To try to cater to local sentiments, the Russians at first used mainly Muslim troops who were related to the Afghan Tajiks, Uzbeks, and Turkomen minorities.

194   *with Pakistani and Saudi Arabian help*   From the Pakistani Directorate of Inter-Service Intelligence (ISI) and the Saudi Arabian external intelligence organization as well as various pious foundations (*awqaf*).

194   *ultimately number about 125,000*   Gérard Chaliand, *Report from Afghanistan* (New York: Viking, 1982, translated from the French *Rapport sur la résistance afghane*), 37–38, and Ahmad and Barnet.

195   *who came from Saudi Arabia*   Who because of his purported skill with

anti-aircraft missiles was known under the nom de guerre "the Archer."

195   *both the Russians and the Americans*   As Peregrine Hodson heard, *"Shuravi dushman e ensaniat, Amrika, dushman e ensaniat"* ("Russia is the enemy of humanity; America is the enemy of humanity"). *Under a Sickle Moon* (London: Hutchinson, 1986), 117. Hodson, who had read Persian at Oxford, almost uniquely among the Western visitors could communicate easily with the insurgents during the trip he described in his book.

195   *usually fewer than thirty men*   A postwar Russian general staff study, quoted by Stephen Turner in *Afghanistan, A Military History* (New York: Da Capo, 2002), 247, called them "small, exceedingly mobile groups of mujtahideen using maneuver tactics."

195   *denouncing the government and the Russians*   Chaliand, 55.

195   *"too spread out to lose it"*   Ahmad and Barnet.

196   *As Barnett R. Rubin then observed*   "The Fragmentation of Afghanistan," *Foreign Affairs*, no. 2 (1989): 153.

196   *quite effective even over long distances*   Hodson, passim. Hodson was able to pass from one area to the next where his hosts would already have been informed of his coming, apparently by runners.

196   *to a given ethnic group*   Three of these were from the Ghilzai Pushtun peoples along the Pakistani frontier and one a Pushtun from the eastern areas of Afghanistan; one was a Tajik, and two claimed Arab ancestry. Neither the exiled king nor the major supporters of the former regime, the Durrani Pushtuns, played significant parts.

197   *told the English visitor Peregrine Hodson*   Hodson, 104.

197   *"We are fighting to protect our religion"*   Hodson, 94.

198   *his reputation for success*   Olivier Roy, *Islam and Resistance in Afghanistan* (Cambridge: Cambridge University Press, 1986), 182.

198   *Then he counterattacked*   Roy, 200–202.

199   *worth listening to their conclusions*   Quoted in John C. Griffiths, *Afghanistan: A History of Conflict* (London: André Deutsch, 1981), 180–181.

199   *for the entire country"*   "Report from Afghanistan," *Foreign Affairs* (Winter 1983–1984): 2, 424.

200   *the most brutal forms of counterinsurgency*   What the Americans did was often reported. For example, see Jonathan Schell in *The New Yorker*, which account was subsequently republished as *The Military Half: An Account of Destruction in Quang Ngai and Quang Tin* (New York: Knopf, 1968). Seventy percent of the 450 hamlets in the province of Quang Ngai were destroyed by "Task Force Orange" and the U.S. Air Force, with annual Vietnamese casualties at between thirty-three thousand and fifty thousand. Sheehan, 686–687. This was comparable to the Russian destruction of the Panjshir valley.

200   *Helsinki Watch summarized their findings*   "Tears, Blood and Cries: Human Rights in Afghanistan Since the Invasion, 1979–1984" (Report: Helsinki Watch, December 1984).

200	*enemy combatants not covered by the Geneva Conventions* Ryan, 154.
201	*As one mujahid told Peregrine Hodson* Hodson, 53.
203	*on a budget in 2007 of $8 billion* Ann Scott Tyson, "New Plans Foresee Fighting Terrorism Beyond War Zones," *Washington Post,* April 23, 2006.
203	*formulas have been announced periodically* The latest, the surge of additional troops, was set forth in President Bush's speech on January 3, 2007. A detailed analysis of it was presented by Anthony H. Cordesman in the *New York Times* of January 12, 2007.
203	*Johns Hopkins School of Public Health* The highly respected British medical journal *The Lancet* had estimated in November 2004 that about 100,000 had died. The new study, made on a much larger sample of Iraqi families in October 2006, raised the figure to between 425,000 and 800,000 more Iraqis having died since March 2003 than would have been expected without the war. While the British government concluded that the study was based on a sound methodology, President Bush dismissed it as not credible. Associated Press, "U.K. Aides Backed Study on Iraq Toll, BBC Reports," March 27, 2007, *International Herald Tribune.*
203	*U.N. High Commissioner for Refugees* Sudarsan Raghavan, "War in Iraq Propelling a Massive Migration," *Washington Post,* February 4, 2007, and Alissa J. Rubin, "Shattered Lives of Baghdad's Sunnis," *International Herald Tribune,* March 27, 2007. "Theirs is a world of ruined buildings, damaged mosques, streets pitted by mortar shells, uncollected trash and so little electricity that many people have abandoned using refrigerators altogether." The report of the International Committee of the Red Cross gave an even more grim account of life in Iraq. Ian Black, "Red Cross Details 'Unbearable Suffering' of Iraqi Civilians," *Guardian,* April 12, 2007.
204	*another one hundred thousand were expected to claim compensation* Scott Shane, "Data Suggests Vast Costs Loom in Disability Claims," *New York Times,* October 11, 2006.
204	*needed mental health treatment* Charles W. Hoge MD et al, on behalf of the Surgeon General, in the Journal of the American Medical Association, March 1, 2006, 295, 1023 ff. Also see Christian Nordqvist, "One Third of Iraq Veterans with Mental Health Problems," *Medical News Today,* March 1, 2006, and Aaron Levin, "Nearly 10 Percent of Iraq Vets Screen Positive for PTSD," *Psychiatric News,* American Psychiatric Association, April 7, 2006.
204	*the rest of their lives* Ronald Glasser, "A Shock Wave of Brain Injuries," *Washington Post,* April 8, 2007.
205	*artillery shells and bombs* I am indebted to Dr. Hans Noll, American Cancer Society Professor of Genetics and Molecular Biology, for the following information: when depleted uranium shells strike their targets, the heat of the impact mutates the uranium into an aerosol of U_3O_8. "It settles as a fine dust, which enters the body in a variety of ways. Uranium

oxide is an extremely potent neurotoxin with a high affinity for DNA. This DNA fragmentation results in genetic defects like cancer and malformation in developing fetuses. The dust particle, because of its high toxicity, is much more harmful than the low radioactivity associated with it. Inhaled as dust, uranium oxide is accumulating in the lungs, liver and kidneys and affect[s] the nervous system within weeks. There is persuasive evidence [that] most of the Gulf War Syndrome is caused by the neurotoxicity of U_3O_8 and not by post-traumatic stress disorder, as claimed by the Pentagon. So far the US government has refused to finance the epidemiological studies to correlate incidents, symptoms and regions of the impact and use of [d]epleted [u]ranium." Personal letter to the author, October 30, 2006.

205   *approximately $500 billion*   Lawrence Lindsey was fired as economic adviser to the president for predicting that the war might cost as much as $200 billion when President Bush was saying that it would be no more than $50 billion. David Leonhardt, "What $1.2 Trillion Can Buy," *New York Times,* January 17, 2007.

205   *between one and two trillion dollars*   National Bureau of Economic Research, Working Paper 12054.

205   *public education here at home*   David Leonhardt, "What $1.2 Trillion Can Buy," *New York Times,* January 17, 2007.

206   *success by any definition*   The "key judgments" of the latest National Intelligence Estimate have been released to the public. They are summarized in the *New York Times*, February 3, 2007. The "intelligence community" concluded that "Unless efforts to reverse these conditions [current trends] show measurable progress . . . we assess that the overall security situation will continue to deteriorate . . ."

206   *than losing power.*   Michael Howard, "U.S. admits Green Zone is no longer safe as suicide bomber strikes at heart of government, *The Guardian*, April 13, 2007, and Ewe McAskill, "Blow to Bush as top U.S. commander warns of worse to come in Iraq, *The Guardian*, April 27, 2007.

206   *it now is self-financing*   According to a U.S. government report, a summary of which was published by John F. Burns and Kirk Semple as "Iraq Rebels Financing Themselves" in the *International Herald Tribune*, November 27, 2006.

206   *black market has surged at least a third in the last year*   C. J. Chivers, "Black-Market Weapon Prices Surge in Iraq Chaos," *New York Times,* December 10, 2006.

207   *the Kurdish area of the north*   Turkish forces began in May 2007 to mass and some actually invaded Iraq, *New York Times*, June 9, 2007.

207   *has pointed out*   published on the internet as AFSATEX@aol.com.

208   *Lieutenant General James F. Amos, USMC*   FM 3–24, MCWP 3–33.5 (Washington, D.C.: Headquarters, Department of the Army, 2006).

208   *for opposing escalation*   On January 4, 2007, President Bush removed the

two senior commanders of American forces in the Middle East, Generals John Abizaid and George Casey, who had opposed the further buildup of American forces, and the director of national intelligence, John Negroponte, who defended his agency's assessment that Iran is "still a number of years off, and probably into the next decade" in acquiring nuclear weapons.

209 *were probably successful* Carnegie Endowment Policy Brief 24, April 24, 2003; also see Pei's article in the *International Herald Tribune*, March 17, 2004.

210 *Hanson Baldwin* New York Times, February 8, 1966.

211 *fury and incomprehension in the Philippines* As James Brooke reported, the more recent campaign in the Philippines against the Muslims who are seeking a separate state or at least regional autonomy recalled the brutal campaign waged on the eve of the First World War by General "Black Jack" Pershing. "The real fear is that this will go back to a Christian-Moro conflict," a priest told Brooke. In the early campaign, the Americans killed about fifteen thousand Muslim Filipinos. "Born Again in the Philippines: The U.S. 'Black Jack' Strategy," *International Herald Tribune*, January 28, 2002. Later in 2002, the U.S. committed about thirteen hundred troops to "advise" the Filipinos against the Moros (the Muslims).

211 *the war will last for years* John F. Burns, "War Could Last Years, Commander Says," *International Herald Tribune*, January 8, 2007.

211 *could be said about the war in Iraq* I have dealt with Iraq in much more detail in *Understanding Iraq* (New York: HarperCollins, 2005 and 2006) and, together with Senator George McGovern, in *Out of Iraq: A Practical Plan for Withdrawal Now* (New York: Simon & Schuster, 2006).

213 *Ahmed Rashid has written* Ahmed Rashid, *Taliban: Islam, Oil and the New Great Game in Central Asia* (London: Tauris, 2000), 21.

213 *came out of religious seminaries* These religious schools are usually called madrassas. The word means simply "place of study" but has come to be applied only to traditional schools teaching literacy and religious subjects. Such schools were supported by nongovernmental foundations and private donations as distinct from modern, secular schools that were supported by government. They were the principal support of the religious revival movements. See Alexander Evans, "Understanding Madrasahs," *Foreign Affairs* (January/February 2006): 9ff.

213 *throughout the Islamic world* Other puritan revivalist movements have included the Sanusi movement in Libya, the Muslim Brotherhood in Egypt, the Mahdiyah in the Sudan, the Salafiyah in Morocco and Algeria, the Ahmadiyah in India, the Jama't-I Islamic in Pakistan, and the Sarekat Islam in Indonesia. These were Sunni movements, but similar attempts to restore "original" Islam also affected the Shiis in Iran and elsewhere.

215 *restore peace throughout their area* Rashid, 25.

215 *rigid, authoritarian, and often ugly order* Many of their practices are less exotic than we think. They are set forth in the Old Testament as well as in

the Quran. Some of the worst, in our eyes, were practiced in the Massachusetts Bay Colony before the American Revolution. Today, apparently, 55 percent of Americans polled by the *Washington Post* and *Newsweek* believe that every word of the Bible is God's commandment and as required by Leviticus 25:25 ("ye shall do my statutes, and keep my judgments, and do them . . .") presumably must (but legally cannot) implement the punishments mandated for various crimes. Since the Old Testament and the Quran essentially agree on these, the Taliban actually execute them. In the Old Testament, a number of crimes are to be punished by stoning to death. Blasphemy is one. (Leviticus 25:14, 16, and 23), apostasy is another (Leviticus 20:2). Working on the Sabbath draws the death penalty. Various sexual transgressions merit burning or stoning to death (Leviticus 18:22; 20:10; 21:9 and 13; Deuteronomy 22:23–24). Many lesser crimes draw public lashings (Deuteronomy 25:2–3), which most of us see as one of the ugliest practices of the Taliban. Finally, and perhaps amusing, is the emphasis the Taliban put on growing great bushy beards. In the aftermath of the attack on the World Trade Center in New York, having a beard got one poor Sikh (not, of course, a Muslim) lynched by an American mob. But Leviticus 19:27 and 21:5, in the curious translation of the King James Version, orders Christians and Jews "Ye shall not round the corners of your heads, neither shalt thou mar the corners of thy beard." In their beards too, the Taliban were following the dictates of the Old Testament. In their dedication in battle and terrorist attacks, the Taliban are encouraged by rewards in the afterlife, as the Catholic priests promised Spanish guerrillas in their fight against Napoleon, as I have mentioned in Chapter Two, that they would be assured of going to heaven "by killing the French heretical dogs," promised that "any soldier wounded fighting the French was ensured 100 years relief from purgatory [and that] anyone killed would be reborn three days later in paradise."

216   *"trying to occupy their country"*   Najibullah Lafraie [Afghan's minister of state for foreign affairs from 1992 to 1996], "The Way Out Is to Get Out," *International Herald Tribune,* September 6, 2006.

216   *"At least they are not thieves"*   "Taliban Rising," *The Nation,* October 30, 2006.

216   *"the police are the main criminals"*   Barnett R. Rubin, "Saving Afghanistan," *Foreign Affairs* (January/February 2007): 60, 67.

216   *"has energized the Taliban"*   Patrick Wintour and Declan Walsh, "UK Has Boosted Taliban, Admits Defence Chief," *The Guardian,* July 8, 2006.

216   *"an elder on the Sperwan village council"*   "NATO's Afghan Struggle," *New York Times,* January 13, 2007.

216   *"seem born every day"*   Charles M. Sennott, "Afghanistan: After 5 Years, a Forgotten War?" *International Herald Tribune,* September 13, 2006.

217   *thirty-five hundred killed and fifteen thousand wounded*   Antonio Maria Costa, "The New Golden Triangle," *International Herald Tribune,* De-

cember 1, 2006. "Iran has deployed almost 20,000 antinarcotic police and border guards along its 1,845 kilometer border with Afghanistan and Pakistan—the world's most active opium smuggling route. Twenty-eight mountain passes have been blocked by huge concrete structures. Hundreds of kilometers of trenches—four meters wide and four meters deep—have been dug to stop drug caravans eluding patrols. Towers and barbed wire stretch as far as the eyes can see."

217   *"Taliban and their Qaeda allies"*   International Herald Tribune, July 24, 2006.

218   *"Islam-is-the answer solution"*   Jeffrey Gettleman, "Somalia on the Brink of Total Collapse," *International Herald Tribune,* February 21, 2007.

218   *to attack Somalia*   Michael R. Gordon and Mark Mazzetti, "U.S. Used Bases in Ethiopia to Hunt Al Qaeda in Africa," *New York Times,* February 23, 2007, and "U.S. Fought Al Qaeda from Ethiopian Soil," *International Herald Tribune,* February 24, 2007.

219   *"by uniformed troops"*   As Sam Kiley points out in "Looting Troops Prey on Somalia's Refugees," *The Guardian,* April 29, 2007, "only the Ethiopian and government forces have uniforms."

219   *"support from neighboring countries"*   Xian Rice, "U.S. Blames Eritreat over Somalian Insurgency," *The Guardian,* April 9, 2007.

219   *"against a foreign occupation has lost"*   Milt Bearden, "Wrong War, Wrong Time," *International Herald Tribune,* February 7, 2007.

219   *"thousands of new anti-U.S. militants and terrorists"*   "In Somalia, a Reckless U.S. Proxy War," *International Herald Tribune,* December 27, 2006.

219   *the society is collapsing*   Jeffrey Gettleman, "Somalia on the Brink of Total Collapse," *International Herald Tribune,* February 21, 2007.

220   *a hundred thousand men and women*   To do so, as Paul von Zielbauer wrote in "Army is Cracking Down on Deserters," *New York Times,* April 9, 2007, "'We're enlisting more dropouts, people with more law violations, lower test scores, more moral issues,' said a senior noncommissioned officer involved in Army personnel and recruiting. 'We're really scraping the bottom of the barrel trying to get people to join.'"

220   *and various parts of Africa*   According to the Department of Defense in 2005, quoted by Chalmers Johnson, *Nemesis: The Last Days of the American Republic* (New York: Holt, 2006).

220   *at the top of the Bush Administration agenda*   Frederick Kagan's December 14, 2006, paper "Choosing Victory: A Plan for Success in Iraq" (www.aei.org/publications/pubID.25292/pub_detil.asp) formed the basis, among other things, for President Bush's speech on Wednesday, January 3, 2007. In that speech, Mr. Bush overrode the advice of the military and even senior members of his own party to adopt Kagan's argument for escalation.

222   *not report such illegal actions*   Thomas E. Ricks and Ann Scot Tyson, "Troops at Odds with Ethics Standards," *Washington Post,* May 5, 2007; Ewan MacAskill, "Iraq War Strain Leads Troops to Abuse Civilians, Sur-

vey Shows," *The Guardian*, May 5, 2007. Violation of the military code is neither uncommon nor just recent, of course. Vietnam offered many examples and even in the early phases of the Iraq war, as a soldier recalled to *New York Times* correspondent Bob Herbert, "Guys in my unit, particularly the younger guys, would drive by in their Humvee and shatter bottles over the heads of Iraqi civilians passing by."

222   *Joseph Lelyveld has written*   New York Review of Books, February 15, 2007.

223   *"humanity hanging from a cross of iron"*   Speech before the National Association of Newspaper Publishers, Washington, D.C., April 16, 1953.

INDEX

Page numbers followed by an *n* indicate a reference in notes.

Abane, Ramdane, 138, 139, 140–41, 244n, 245n
Abbas, Farhat, 133
Abdul-Qadir, 128–29, 130, 134, 138
Abizaid, John, 256n
Acheson, Dean, 159, 166
Adams, Gerry, 70, 71
Adams, John, 6
Adams, Samuel, 227n
Aden, 122
Afghanistan, xxiii, 181–201; Afghan casualties, 201; "after-war," 19; American intervention, xxvi, 190, 253n; American proxy war in, 194, 196; Americans in, hatred of, 216–17; American war in, 202, 212–17; beneficiaries of American arms in, 212; bin Laden and, 191, 195, 215, 253n; Britain and, 98, 184–90; communications, 196; counterinsurgency in, 31; Daoud's republic, 193–94; drug trade in, 212–13, 217, 258n; external and internal resistance, 196, 254n; First Afghan War, 186; flight of inhabitants to Pakistan, 194–95, 196, 201; Fourth Afghan War, 195; guerrillas or mujahideen, 192, 195, 254n; guerrilla warfare in, xviii, 187, 192, 194–201, 252n; imperatives of revenge and hospitality, 191; insurgents, attacking of fellow inhabitants, 210; insurgency against British, 186–87, 252n; Islam in, 186, 191–92, 195; Khyber Pass, 187; lack of unity among resistance, xxiii, 195, 197; land and its inhabitants, 182, 190–92; military technology and terrain, 86; monarchy of, 190, 193, 253n; Northern Alliance and Massoud, 197–98; as part of world Islamic revolution, 201; Polk in, 190–91, 252n; Qandahar and beginnings of modern Afghanistan, 252n; Russian withdrawal, 211; Russian casualties, 199; Russian counterinsurgency, 200–201, 254n; Russia in, 181–86, 189, 193–201, 252n, 253n; Russo-Britain rivalry over Central Asia and, 184–85; Sayyaf government-in-exile, 196; Second Afghan War, 188–89; Shah Shuja regime, 186; size of insurgent movement, 96, 254n; social organization and culture, 191–93, 197, 253n; Taliban in, xxiii, 201, 213–17, 258n; Third Afghan war, 190; Wakkan Corridor, 189–90; warlords, 193, 212–13; WWI and, 190; xenophobia in, 182, 186, 197, 209
Aguinaldo, Emilio, 39, 40, 41, 42–43, 230n
Ahmad, Eqbal, 195, 225n

Algeria, xv, 19, 124–43; American supplied war matériel, 243n; battle of Algiers, 140, 142, 245n; Ben Bella leadership, xxvi, 135, 139, 142, 229n; climate
of insurgency, 34; Communist Party
in, 135–36, 140; concentration camps,
143; conventional warfare, 139; cost of
war, 122, 142–43; counterinsurgency,
31; creation of alternative administration, 138–39, 198; effect of Dien Bien
Phu, 136, 137; empty promises of
freedom for, 134; European settlement,
130–31; European view of people,
132; Evian Accords, 142; execution of
terrorists, 122; FLN, 136–43, 244n,
245n; French army recruits, 155;
French brutality, 24, 34; French Communist Party and, 134–35; French
conquest and occupation, 124–25,
127–28, 242n; French counterinsurgency, 136, 137–38; French destruction of indigenous culture, 131, 136,
154, 244n; French forces, size of, xx,
137; French use of torture, 141, 143,
245n; growth of insurgency, xx, 137;
guerrilla warfare, xviii, 125, 128–29;
insurgent leaders, 138, 139; insurgents,
attacking of fellow inhabitants, 138;
insurgents' "nation-in-motion," 129;
insurgents, number of, 33, 96, 135,
137, 139, 244n; internal resistance
forces, 86; Islam in, 132; Jamiyah,
132–33, 243n; Kabyle people, 138,
244n; loss of popular support and
defeat of Abdul-Qadir, 130; Messali
Hadj leadership, 133–34, 135, 140,
155, 243n; Morice Line, 142; Mouvement pour le Triomphe des Libertés
Démocratiques, 135; national government, 138–39; nationalism, 128, 134,
135, 136, 137, 140, 242n, 243n; native
suppression of guerrillas after independence, xxvi, 142, 143, 229n; OAS,
141–42, 245n; OS, 135; Ottoman
Turks ruling, 126–27; Parti Progressiste Algérien, 133–34; peoples of, 126;
Philippeville massacre, 137, 140; physical and social geography, 125–27; Polk
and, xv, 131, 243n; public opinion and,

68; resistance movements, 128–29,
132–34, 135, 136, 138, 243n; Secret
Army Organization, 70; terrorism in,
138, 141–42; Unionists, 133; war for
independence, 137–43, 244n, 245n;
WWII and, 134
Amanullah, King, 190, 193, 253n
America: in Afghanistan, xxiii, xxvi, 190,
212–17, 253n; al-Qaida attacks on,
215; armed forces, demoralization
and violations of standards, 220, 222,
259n; arming of French in Algeria, xv,
137; Cold War and, 98, 164, 194; cost
of "long war," 203, 222–23; costs of
Iraq war, 205, 255n, 256n; counterinsurgency in Greece and, 105–6; fear
of Communism, 50, 164, 165; foreign
military and air bases, 220; Gulf
War (1991), 205, 255n; Hawaii taken
by, 39; Indian warfare in (see Native
Americans); Iran, plans for war against,
202, 220, 256n; in Iraq, xviii, xxvi,
122, 128, 201, 203–11; liberation of
Europe, WWII, 100; neoconservative
plan for Iraq, 207–11; neoconservative
plans for "long war," 202–3; news of
war blocked, 43–44, 230n; in the Philippines, xx, 35, 40–53, 202, 230n; plan
to impose democracy on other nations,
33; proxy wars, 194, 202, 218–19;
Roosevelt's anti-imperialist policy, 157;
SEATO, 170; series of wars against
insurgencies, 202; in Somalia, xxvi,
202, 218–19, 222; Spanish-American
War, 39–40; Truman Doctrine, 105,
166; unification of Europe against the
USSR, 165; in Vietnam, xv–xvii, 35,
43, 50, 122, 136, 150, 163–80; Vietnam, support for French, 159–62
American Enterprise Institute, 207
American Revolution, 1–19, 227n; battle of
Bunker Hill, 8; battle of Kettle Creek,
13; battle of Moore's Creek Bridge,
12, 228n; Braddock's defeat and, 4, 5,
136, 158; British soldiers in, 2; climate
of insurgency, 5–11; "Committees of
Safety" as alternative administration,
xxii, 12–13, 174; Continental Line,
xxv, 4, 16, 18, 19, 226n; control of

territory and, 13, 227n; counterinsurgency in, 12–13; The Cowpens, 228n; creation of permanently held rebel-controlled areas, 15; disruption of British rule and, 6–7; establishing pattern of insurgency, 14; flight of Loyalists, 14; French forces in, 2, 18–19; French-Indian allies, 2; guerrilla warfare in, 2–3, 4–5, 11, 228n; "Irregulars" as guerrilla fighters, 11; lack of supplies and ammunition, 17; General Charles Lee's approach, xxv–xxvi, 15, 16, 227n; Loyalists in, 7–8, 10, 11, 12, 13, 14, 30, 31, 210, 227n; military leaders, 11, 18; nationalism, 11; Phase 1, 6, 8, 12; Phase 2, 17–18; Phase 3, 11, 18–19; repressive British policies and, 8–10; superiority of British force, 1, 69; terrain and, 2; threat as recruiting incentive, 13–14; Washington's approach, xxv, 4–5, 15, 16, 18–19, 229n

Amin, Hafizullah, 193–201

Amos, James F., 208, 210, 211

Anderson, David, 111, 115, 122, 240n

Arnold, Terrell, 207

Art of War, The (Jomini), 33, 41

Askwith, Thomas, 120–21

Baldwin, Hanson, 210

Bao Dai, 148, 159, 166, 169, 176

Baring, Sir Evelyn, 115, 116–17, 118, 121, 241n

Barnet, Richard, 195, 225n

Battle of Algiers, The (film), 245n

Begin, Menachem, xix

Ben Bella, Ahmad, xxvi, 135, 139, 142, 229n, 244n

Best and the Brightest, The (Halberstam), xvii

Bilmes, Linda, 205

bin Laden, Osama, 191, 195, 215–16, 253n

Bismarck, Prince Otto von, 107

Black Hawk Down (film), 218, 219

Blunt, Wilfred, 127, 128, 242n

Bonifacio, Andrés, 38

Boumedienne, Houari, 136

Bouquet, Henry, ix, 2

Bowles, Chester, 190

Bowring, Philip, 53

Braddock, Edward, 4, 5, 136, 227n

Braudel, Fernand, 22

Brezhnev, Leonid, 182

Bright and Shining Lie, A (Sheehan), xvii, 246n

Britain: in Afghanistan, 216; African empire, 189 (*see also specific countries*); American Revolution and, 1–19, 228n; Anglo-Spanish campaign against Napoleon, xvii–xviii; Braddock's defeat, 4, 5, 136, 158, 187; Cairo Conference, 232n; campaigns against Native Americans, 56, 57; in China and Hong Kong, 149–50; combat tactics, 1–2; in Cyprus, xix; Great Game, 98, 181, 182, 185, 190, 193; in Greece, 48, 90–91, 97–103, 234n , 239n; Indian empire, 98, 107, 182–83, 187–88, 190; in Ireland, 44, 55–71; in Kenya, xxiii, 68, 107–23; Khyber Pass, 187; in Malaya, xxiii, 251n; military strength, eighteenth century, 1; in Pakistan, 183; Palestine Mandate and, xix, 62; pillaging by army, 2; pirates, 55; policy in Mediterranean and Middle East, 98; response to IRA bombing campaign, 70; in Rhodesia, 68; Russo-Britain rivalry over Central Asia, 184–85, 189–90; in South Africa, ix, 44; Spain and, 55–56; support of French in Indochina, 164; Tehran Conference, 88, 238n; Tito's Partisans and, 83–85, 86, 88; use of mercenaries, 1; war as theater, 3. *See also* Ireland

Brooke, James, 257n

Bugeaud, Thomas, 129

Burgoyne, John, 13

Burma, 165; guerrilla warfare in, xviii

Bush, George W., 50; American casualties and costs in Iraq, 204–5, 255n, 256n; cost of "long war," 203, 222–23; lack of congressional oversight, 203; neoconservative plan for Iraq, 207–11; neoconservative plans for "long war," 202–3, 220–23; new legal arrangements, 221–22; new Middle East goals, 207; plans for Somalia, 219; preemptive warfare, 203; removal of

Bush, George W. (*cont.*)
commanders, 256n; Special Operations
Command, 202–3
Buttinger, Joseph, 154–55

Caesar, Julius, 31
Callwell, Charles E., 42, 44
Camus, Albert, 143
Canh, Prince of Vietnam, 148, 149
Carlos IV, King, 21, 228n
Carlos V, King, 21
Carver, George, xxii, 174
Casey, George, 256n
Castro, Fidel, xx, xxvi
Catherine the Great, 185
Central Intelligence Agency (CIA), xv, 97;
in the Philippines, 50; in Vietnam,
xxii, 157, 163, 168, 169, 174, 179,
241n, 245n
Chaliand, Gérard, ix
Chamberlain, Joseph, 240n
Chartrand, René, 29
Chiang Kai-shek, 164
China: British in Hong Kong, 149–50;
Mao's guerrillas in, 86; size of insur-
gent movement, 96; small-scale arms
industry, 29; Vietnam and, 146, 148,
151, 160. *See also* Mao Tse-Tung
Chou En-lai, 168
Churchill, Winston, 78, 91, 101, 134, 164,
183, 232n, 233n, 235n; Cold War and,
98; deal with Stalin, 100–101, 239n;
domino theory and, 159; Greece and,
97–103, 237n, 238n; Greek resistance
as bandits, 31, 99, 238n
Clarke, Elijah, 11, 18
Cleveland, Grover, 39
Clinton, Bill, 218
Clive, Nigel, 239n
Colby, William, 136, 168
Collins, Michael, 60
Collwell, Sir Charles, 128–29, 242n
Colombia: guerrilla warfare in, xviii
Communism: American fear of, 50, 164,
165; in the Balkans, 99; domino
theory, 99, 159, 164, 166, 252n; in
Europe, 99; in France, 134–35, 159,
248n; KKE in WWII Greece, 94, 95,

104, 106, 166; in the Philippines, 49;
Tito and, 81–82, 88, 98, 101; in Viet-
nam, 151. *See also* Stalin, Josef
Cooper, Chester, 225n
Cornwallis, Charles, 13, 18, 227n
counterinsurgency, xxiii; efficacy of, 35,
45; enclave strategy, 210; failures of,
31; meet danger with force mentality,
117. *See also specific countries*
Counterinsurgency Field Manual, 207–11
Crary, Catherine S., 12
Cuba, xviii, xx, xxvi, 38, 39–40
Curzon, Lord George, 190
Cyprus, xviii, xix, xx, 122, 236n

Daoud, Sadar Muhammad, 193–94
Davidson, Basil, 86, 235n
Davie, William, 11
Deakin, F. W. D. (William), 81, 83, 85,
88, 235n, 238n, 244n
de Gaulle, Charles, 134, 141–42, 159,
237n, 245n
de Sola Pool, Ithiel, 225n
de Valera, Éamon, xxvi, 60, 62, 63, 65,
142, 229n
Dewey, George, 40, 230n
Diem, Ngo Dinh, 168, 169–71
Djilas, Milovan, 102
Draper, Theodore, 225n
Dulles, John Foster, 50, 167, 169, 170, 182,
252n
Dupree, Louis, 186, 252n

Eden, Anthony, 91, 100, 101, 102, 237n,
238n
Egypt: abortive invasion of, 1956, 141;
Britain and, 98; Greece in exile in,
92, 99, 103; Islam groups in, 132;
Napoleon in, 98, 127, 183; World War
II and, 92, 96–97, 99
Einaudi, Luigi, 225n
Eisenhower, Dwight D., 100, 166, 167,
190; on cost of war, 223
Elizabeth I, Queen, 56, 183–84
Elkins, Caroline, 115
Ellsberg, Daniel, 225n, 246n
England. *See* Britain
Equaguirre, Andrés, 31

Erskine, George, 117, 118
Escape to Adventure (MacLean), 234n
Ethiopia, 218

Fairbank, John King, 225n
Fall, Bernard, xxi, 173–74, 248n
Fanon, Franz, 244n
Feleo, Juan, 49
Fernando, Prince, 21, 22–24, 32–33
Fire in the Lake (FitzGerald), 172
FitzGerald, Frances, 172, 225n
France: in Algeria, xv, xx, 122, 124–43,
    242n; American backing in Vietnam,
    164–67, 243n; American Revolu-
    tion and, 2, 18–19; Anglo-Spanish
    campaign against Napoleon, xvii–xviii;
    battle of Fontenoy, 4; Communist
    Party in, 134–35, 159, 248n; Dien Bien
    Phu, 136, 137, 157, 162, 167, 169, 187;
    Evian Accords, 142; in Madagascar,
    107; resistance movement, World War
    II, 97–98, 237n; size of military,1;
    Spain's guerrilla war against, 20–34;
    the *tache d'huile* (oil spot), 15–16, 210;
    use of *la petite guerre*, 3–4, 27; Vichy
    government, 134, 157, 237n; in Viet-
    nam, xv, 149–62, 170; war as theater, 3
Franklin, Benjamin, 5, 6, 187–88
Fraser, Hugh, 117
Frazer, Jendayi, 219, 239n
Frederika, Queen, 99, 238n
Frist, Bill, 217

Gage, Thomas, 12
Gall, Carlotta, 216
Gallieni, Joseph, xxii, 153, 154, 160
Gandhi, Mohandas, 99
George, David Lloyd, 17, 31, 60
Germany: alliance with Japan, 92; in East
    Africa, 107; in Greece 94, 237n; Nazis
    in Yugoslavia, xx, 73–77, 83; war as
    theater, 3. *See also* World War II
Ghafar, Hajji Abdul, 216
Giap, Vo Nguyen, xxv, 157, 169, 180,
    210–11, 229n, 246n, 248n
Gorchakov, A. M., 188
Goya, Francisco, 24
Granada, 209

Gravel, Mike, 246n
Greece, 73, 89–106; American advisors
    in, 105, 157; Britain and, 48, 90–91,
    97–103, 234n , 239n; Churchill-Stalin
    deal and, 102–3; the Church in, 95;
    civil war, 104–5; collaboration in,
    93, 96, 97, 210; Communist Party in
    (KKE), xxii, 94, 95, 103, 104, 106,
    166, 176, 237n; creation of alterna-
    tive administration, xxii, 174; creation
    of permanently held rebel-controlled
    areas, 15, 137; domino theory and,
    166; EAM/ELAS insurgency, xix,
    xxii, xxiv, 17, 31, 44, 94–106, 137,
    139, 140, 161, 174; EDES, 97, 197;
    exchange of populations with Turkey,
    68; fascism in, 90–91, 92–93, 236n;
    foreign aid and insurgents, 97; "Free
    Greece" formed, 94; German occupa-
    tion, 94, 237n; growth of insurgency
    in, 94; in-fighting and destruction of
    insurgents, 39, 44, 176; lack of leader-
    ship for insurgency, 95; loss of popular
    support and defeat of insurgents,
    104–6; Metaxas leadership, 90–91,
    92–93, 94, 95, 96–97, 99, 236n,
    237n, 239n; monarchy and royalist
    government, 89–90, 92, 96–97, 99,
    103, 104–5, 237n; nationalism, 90, 93,
    106; number of guerrillas, 33, 94, 96;
    PEEA, provisional government, 1944,
    99; political history, 89; Polk in, 104;
    right-wing terrorist groups (X), 93,
    103, 236n, 239n; "rule of the Colo-
    nels," 239n; terrorism in, xix; village
    organization of insurgency, 94, 96
Greene, Nathanael, 18, 228n
Griffith, Arthur, 60
Griffith, Samuel B., x
Grivas, George, 236n
Guantánamo, 233n, 241n
guerrilla warfare, xiv, xvi, xviii, 188;
    administration, xvi; becomes superflu-
    ous once insurgents win, xxvi, 32–33,
    62–65, 65, 142, 143, 229n; Che Gue-
    vara on, ix–x; in China, 86; control of
    territory and, xxiii, 13; cost in lives and
    money, xviii; defined, xvii–xviii, 3, 26,

guerrilla warfare (*cont.*)
228n; eyewitnesses to, 84–85; fighters in, x, 27, 95–96; foreign aid and, 86, 97, 239n; ideas matter more than material resources, 87–88; isolation and, 105; longest in modern history, 180; loss of popular support and defeat of, 103–4, 104–5; Mao Tse-Tung on, x; as "national wars," 33, 209; in Phase 3 (combat), xxiv–xxvi; politics, xvi; relationship to local community and, x; "sea" (citizenry) and "fish" (militants), 33–34, 168; small numbers of insurgents and, xix, 29–30, 94, 111, 137, 157; study of, ix, x; the *tache d'huile* (oil spot), xxii–xxiii, 15, 153; as tactic of insurgents, xiii; tactics, 2–3; T. E. Lawrence on, 242n; terrain and, 2, 27, 85–86, 236n; "the war of the flea," xxiv, xxv–xxvi, 27, 69; treatment of prisoners, 32, 45; as "wars of opinion," 33. *See also* insurgency; *specific countries*
Guevara, Che, ix–x

Halberstam, David, xvii
Hamas, 133
Hancock, John, 6
Harkin, Paul, 178
Haynes, Fred, 225n
Hegel, Georg Wilhelm, 219
Helsinki Watch, 200
Henry II, King, 55
Henry VIII, King, 55
Hitler, Adolf, 74, 82, 90, 92, 234n, 237n
Hizbullah, 133
Ho Chi Minh, xvi, 29, 30, 50, 148, 155, 159, 246n, 250n; Americans and, 157, 158–59, 165; as Asian Tito, 166; on the Chinese, 146; Communism and, 166; the French and, 146, 165; political and administrative approaches, 161; popular support for, xx, 173; resistance to Japanese, 156–58; wins election, 166
Hodson, Peregrine, 197, 201, 253n
Hoffman, Stanley, 225n
Horne, Alistair, 243n
Howe, William, 13
Hudson, William, 234n
Hull, Cordell, 100, 238n
Human Rights Watch, 222–23

Huntington, Samuel P., 225n
Huxley, Elspeth, 111

India: British Empire and, 98–99, 164; British massacre at Amritsar, 116; Gandhi's independence movement, 99; Iron Curtain and, 100; "Mutiny," 187–88; Northwest Frontier, 183, 190; Russian threat to, 194; Russo-Britain rivalry over Central Asia and, 185
insurgency: acquisition of goods and resources, 29–30; alternative administration or "anti-state," creating, xxii, 17–18, 86–88; American wars against, series of, 202; attacking of fellow inhabitants and, 14, 30, 138, 173–74, 210; as bandits, 12, 17, 27, 31, 152–53, 238n; broken people cannot revolt, 58–59; the climate of insurgency, creating, xiv, xviii, 5–11, 24–26, 34, 38–39; creation of permanently held rebel-controlled areas, 15, 86; efficacy of counterinsurgency and, 35; evaluating victory, short term vs. long term, 35; fundamental motivation in all, xiv–xv; growth of, as response to dominant government's repression, xx–xxi, 29–30, 34; guerrilla warfare, xiii (*see also* guerrilla warfare); imperial or colonial power and, xiv, 5–11, 31; in-fighting and destruction of, 39, 44–45, 105–6; initiation into group, 112; labeled as bandits, 16–17; loss of popular support and defeat of, 45, 105–6; Mao's "sea" (citizenry) and "fish" (militants), 10, 44–45; Mao Tse-Tung on, 10; military aspects, xviii; mistake of conventional warfare, 18, 39, 139, 162, 198, 229n; operative society, xiv; opposition to foreigners and, xiii, xiv, xv, 209, 210; outbreak of violence, xviii; Phase 1 (political), xix–xxi, 143, 173; Phase 2 (disruption of dominant power's government), xxi–xxiv, 6, 17–18, 69, 143, 173–74; Phase 3 (combat), xxiv–xxvi, 143, 174, 174–75; as politics of last resort, xiv; revolutionary "cell" organization, 38; securing the political element, 15; small numbers of insurgents, 94; stages

of, xviii–xxvi, 143, 173–74; territoriality and, xiii; terrorism phase, xiii, xviii, xix–xxi; threat as recruiting incentive, 13–14; what happens if they lose, xxvi; what happens if they win, xxvi, 19, 32–33, 142. *See also* guerrilla warfare; *specific countries*

In the Midst of Wars (Lansdale), 51

IRA (Irish Republican Army): bombing campaign in England, 69–70; called "bandits," 17, 31, 60; Continuity IRA, 66, 233n; de Valera declares illegal, 63; failure, xxi; formation of, 59, 60; handbook, 54, 69; negotiated settlement, 70–71; Provos, 67, 68, 70, 71; Real IRA, 71; recruitment by, 61, 232n; Sinn Féin, 60, 62, 63, 65, 71; split into Official IRA (ORIA) and Provos (Provisional IRA), 65–66; terrorism, 60, 69–70; "the war of the flea," 69; in Ulster, 63–71

Iran, 98, 100, 196–97; American invasion plans, 202, 220

Iraq: American counterinsurgency, 208–11; American war in, xviii, xxvi, 25, 44, 122, 128, 201, 203–11; American war in, cost, 122, 203–4, 205, 255n, 256n; Britain and, 98; civil war, 206–7; clear and hold strategy, 105, 210–11; control of territory and, 227n; *Counterinsurgency Field Manual* and neoconservative plan for, 207–11; embedding of reporters, 43; Green Zone, 187, 206, 256n; insurgents attack on fellow inhabitants, 14; National Guard called up, 43; number of insurgents, 33–34; predictions for, 207, 211, 256n

Ireland, 54–71; after-war, 19, 71; attacks on Royal Irish Constabulary, 61; "Bloody Sunday," 67; Bobby Sands, death of, 70, 233n; Dail Eireann, 60; Dail Uladh, 67; Declaratory Act, 57–58; de Valera and suppression of insurgency, 62–63, 65, 142, 229n; Easter 1916, 59; Eire Nua, 67; English attempt to break cultural traditions, 57, 232n; English Black and Tans, 61; English colonialism, 57; English policy of removal, 56; English repression and brutality, 67; English

response to IRA bombing campaign, 70; English sanctions, 57; English "shock companies," 44, 61; Englysshe Pale, 55, 62; first Anglo-Irish war, 60; formation of Free State and Ulster, 62, 64, 232n; genocide in, 56; Good Friday Agreement, 70, 71; guerrilla warfare in, xviii, 60–61 (*see also* IRA); hanging of Pearse, 59; hunger strikes, 70, 233n; idealized shared memory, 8, 58; IRA (*see* IRA); IRB, 60; nationalism, 59; negotiations to end conflict, 67–68, 70–71; Norman invasion, 54–55; "obedient shires," 55; potato famines, 58; price of insurgency, 65; Protestant-Catholic troubles, 63–71; Protestant terrorist campaign, 70; Protestant Unionists, 71; public opinion, 68; recognition of independence, 62–63; Roman Catholic Church in, 58; Royal Ulster Constabulary, 64, 66, 68, 71; size of English forces, 68; size of insurgent movement, 68, 96; Spanish alliance, 55–56; Stormont, 62, 67; terrorism, 60, 64–65, 66–67, 69, 70; Ulster, problem of, 57, 59, 63–71, 232n; Ulster Defence Regiment, 66, 68; Ulster Volunteer Force, 59

Islam. *See* Muslims

Israel: barrier to fence off Palestinians, 68; Haganah, xix; Irgun Zeva'i Le'umi, xix; Palestine Mandate and, xix, 62, 233n; Stern or LEHI, xix; terrorism, xix. *See also* Palestine

Itote, Waruhiu, "General China," 115, 118–19, 123

Ivan IV, "the Terrible," 183

Jackson, Andrew, 11, 19
James I, King, 57, 62
Johnson, Lyndon, 179, 180
Jomini, Antoine Henri, ix, 33, 34, 41–42, 229m
Jordan: Britain and, 98

Kagan, Frederick, 207
Kahin, George, 225n
Kariuki, Josiah Mwangi, 112
Karnow, Stanley, 153, 157, 246n, 248n
Kasper, Sara, 209

Kazan, Khanate of, 183
Kennedy, John F., 50, 175
Kenya, 107–23, 218, 219; attacking of fellow inhabitants and, 210; British forces in, 116; British massacre at Kiruara, 116; British plan of rehabilitation, 117–18; British suppression, 108–9, 115–16; British withdrawal, 68; Christianity in, 122; concentration camps, xxiii, 119–20, 241n; cost, 122–23; Emergency (1952), 115; European settlement of, 63, 108–9, 114–15, 240n; execution of terrorists, 122; "General China," 115, 118–19, 120, 123; Kenya African Union, 110–11; Kenyatta leadership, 111, 115–16, 123, 241n; Kiambaa Parliament, 111; Kikuyu, 107–11, 115, 117, 122, 123, 240n (see also Mau Mau); King's African Rifles, 115, 116; kinship and tribal groups, 108; land reform scheme, 120–21; Malaya plan for counterinsurgency, 118–20; Masai, 108, 109, 239n; Mau Mau, xxiv, 111–22; national independence, 123; peoples of, 8, 108, 110, 240n; politics lost the war, 122–23; resettlement, 110, 172, 210; size of guerrilla forces, 119; small number of casualties, 112; World War II and, 110
Kenyatta, Jomo, 111, 115–16, 123, 240n, 241n
Kessler, Richard, 49, 230n
Kipling, Rudyard, ix, 41
Kissinger, Henry, 225n
Komer, Robert, 136
Korean War, 167, 252n
Krim, Belkacem, 138
Kuwait: Britain and, 98

Laber, Jeri, 200
Lanessan, Jean Marie de, 152
Lansdale, Edward, 121, 148, 153, 169, 170, 231n
Lattre de Tassigny, Jean de, 162, 248n
Lawrence, T. E., 242n
Leakey, Louis, 121, 122
Lebanon: Shia Hizbullah, 133
Lee, Charles, xxv–xxvi, 12, 15, 16, 227n
Lelyveld, Joseph, 222

Lindsey, Lawrence, 256n
Lone, Salim, 219
Louis XV, King, 21
Louis XVI, King, 148
Lyautey, Louis, 153

MacArthur, Arthur, 48
MacArthur, Douglas, 48, 49
Machiavelli, Niccolò, 50, 169
MacLean, Sir Fitzroy, 78–80, 85, 87–88, 100, 101, 234n
MacLeod, Iain, 123
Macmillan, Herald, 102, 123
Macnaghten, Sir William, 186
Madagascar, 15–16, 107
Magsaysay, Ramón, 50–52, 148, 169, 231n
    See also Philippines
Malaya: attempt to isolate people from rebels, xxiii, 172, 210; execution of terrorists, 122; model for Vietnam, 118, 241n, 251n
Malaysia: guerrilla warfare in, xviii
Malhuret, Claude, 199
Mao Tse-Tung: "clear and hold" strategy, 210–11; on control of territory, xxiii; conventional warfare by, 162; guerrilla forces of, 29, 86; on guerrilla warfare, x, xxv, 152; on identifying with the people, xx; population determines success, 10; "sea" (citizenry) and "fish" (militants), 10, 33, 44, 168, 196
Maria Luisa, Queen, 21
Marion, Francis, 11, 18
Mary I, Queen, 56
Massoud, Ahmad Shah, 197–98
Massu, Jacques, 136, 141, 244n
Mau Maus, 111–22, 240n; manufacture of firearms, xxiv, 119; "oathing," 111, 112–13, 114; plan to release from their oaths, 120, 121–22, 153; reasons to revolt, 114–15; refusal to give up, 120–21; subdued by Kenyatta, 123
McAlister, John, 160–61
McDermott, John, 225n
McKinley, William, 39, 41, 43
Medicines sans Frontiers, 199
Merrill, Gyles, 231n
Messali Hadj, 133–34, 140, 155, 243n

Metaxas, John, 90–91, 92–93, 94, 95, 96–97, 99, 170, 236n, 237n, 239n
Mihailovic, Drazha, xix, xx, 77, 88, 97, 101, 140, 234n
Mina, Javier, 28–29, 29, 30, 32
Minh-Mang, 149
Mitchell, George, 70
Mohammaed, Dost, Afghan king, 186
Molotov, Vyacheslav, 100, 101, 238n
Morgenthau, Hans J., 225n
Mountbatten, Louis, 1st Earl, 69
Moyne, Lord, 233n
Murat, Joachim, 22
Murray, J. C., 105
Mus, Paul, 161
Muslims, 196; in Afghanistan, 191–92, 195, 196; insurgency and, xviii; in Iraq, 204; jihad, 192, 215; madrassa, 213, 257n; mullahs, 192; in the Philippines, 53, 196, 257n; puritan or fundamentalism, 213, 215, 257n; Shia, 196–97; in Somalia, 218; Sunni, 196, 204, 207; Taliban, xxiii, 201, 213–17; world Islamic revolution, 201
Mussolini, Benito, 74, 90, 92, 93, 234n
Myers, E. C. W., 237n

Nadir Khan, King, 193
Napoleon, xviii, 16–17, 98; desire for India, 149; in Egypt, 98, 127, 183; relations with Algeria, 127–28; Spain's guerrilla war, xx, 20–34, 149, 209
Napoleon III, 130, 150
Nasser, Gamal Abd al-, 136
"National Defense Strategy of the United States of America," 221–22
National War College: Department of International Studies, 207–8; Polk speaks on Vietnam at, xvi–xvii, 250n
Native Americans: Britain and, 8–9, 56, 108; ceremony of war, 3; European confiscation of lands, 130; fighting style, ix; guerrilla tactics, 3; Pontiac's rebellion, 8; warriors, 2
Navarre, Henri, 162
Nguyen Anh, Emperor, 148, 149
Noll, Hans, 255n
No More Vietnams, xvii, 226n
Nordeau, Max, 240n

Odierno, Raymond T., 211
Omar, Mohammed, 215
Orwell, George, 222
Otis, Ewell, 43–44

Paisley, Ian, 71
Pakistan: Afghan refugees in, 194–95, 196, 201; American proxy war in Afghanistan and, 194, 196; Britain and, 183; Russian threat to, 194
Palestine: execution of terrorists, 122; Mandate, xix, 62, 233n, 240n; public opinion and, 68; Sunni Hamas, 133; Zionist attacks on Britain, 233n
Panama, 209
Pancake, John, 10, 12–13
Parenti, Christian, 216
Partisan Picture (Davidson), 235n
Pearse, Padhraic, 59, 60
Pei, Minxin, 209
Pentagon Papers, The, xv, xxiii, 158, 164–65, 166, 171, 175, 176, 178, 246n, 248n, 249n
Pershing, "Black Jack," 257n
Peter the Great, Tsar of Russia, 184–85
la petite guerre, 3, 11–12, 27, 85, 235n, 251n
Petraeus, David H., 208, 210, 211
Pfeffer, Richard, xvii, 225n
Phan Dinh Phung, 152, 153
Philippines: after-war, 19; Aguinaldo and, 39, 40, 41, 42–43; American forces, size, 44; American intervention in, 40–53, 230n; American Philippines Commission, 41; American proxy war, 202; American repression, xx, 42–44; American view of native people, 41, 42, 230n; atrocities in, 43; Bonifacio and, 38–39; "bought" by America, 41; casualties, 45; Catholic Church and, 36, 39; colonialism of Spain and, 36–37; counterinsurgency, 45, 49, 105; creation of permanently held rebel-controlled areas, 15; EDCOR and land reform, 52; Far East Squadron 311, 47–48; genocide in, 36; government corruption, 51; guerrilla warfare in, xviii, 42–44, 86, 210; Hukbalahap (Huks), 47–48, 49, 51, 52, 53, 169,

Philippines: after-war (*cont.*)
230n; *illustrados*, 37, 155; independence
recognized, 48, 231n; in-fighting and
destruction of insurgents, 39; insurgency, size of, 49; insurgency in, 35–53;
insurgents as bandits, 31; Japanese in,
46–47; Katipunan, 38–39; KPMP, 47;
La Liga Filipina, 37–38; Lansdale and
Magsaysay, 50–52, 169, 231n; as model
for Vietnam, 35, 172; Muslim movements, 35, 53, 196, 257n; news block,
43–44, 230n; Philippine Constabulary,
46, 47; physical and social geography,
35–36; PKP, 49; post-World War II civil
war, 35, 49; recruitment of native army,
44, 51; resistance to American forces,
42–44; resistance to Japanese, 47–48;
Rizal and martyrdom, 37–38; Roxas
and, 48–49, 231n; Spanish reprisals,
xx, 38; Taft's civic action program, 45,
46, 51, 52, 231n; terrorism, 43; uprising
against the Spanish colonials, 35–41;
World War II, 48, 231n
Pickens, Andrew, 11, 18
Pigneau, Pierre-Joseph-Georges, 148–49
Pijade, Mosa, 86
Pike, Douglas, 145
Polk, Thomas, 227n
Polo, Marco, 147
Porter, Roy, 57

al-Qaida, 215–16

Rallis, Ioannis, 93
Rashid, Ahmed, 212–13, 214
Reilly, John, 225n
Reischauer, Edwin, 225n
Rhodesia, 63, 68, 114
Rizal, José, 37
Roosevelt, Franklin D., 100, 134, 157,
248n; Atlantic Charter, 165
Rostow, Walt, xvi, 250n
Rove, Karl, 50
Roxas, Manuel, 48, 231n
Roy, Olivier, 198
Royster, Charles, 4
Rubin, Barnett, 200
Rudforff, Raymond, 21
Rumsfeld, Donald, 202–3, 221

Russia: in Afghanistan, xxiii, 86, 181–201;
Çeçnya, xviii, 201; Cold War, 98, 164;
domino theory, 164; "forward" policy,
194; Greece and, 99, 106, 238n; imperialism of Tsarist, 183–86, 188; Iron
Curtain and, 100, 238n; partisans,
85; *la petite guerre* used by, 85, 235n;
Russian Revolution, 81; Russo-Britain
rivalry over Central Asia, 184–85,
189–90

Sands, Bobby, 70, 233n
Saudi Arabia, 194, 196
Sayyaf, Abdur-Rabb, 196
Schaw, Janet, 227n
Schlesinger, Arthur, Jr., 225n
Scowcroft, Brent, 205
Shay, John, 11
Sheehan, Neil, xvii, 169–70, 175, 178,
246n
Shy, John, 7
Singh, Ranjit, 183
Smuts, Jan Christiaan, 99, 159, 238n
Somalia, xviii, xxvi, 218–19, 222, 239n;
American proxy war in, 202, 218–19
South Africa, ix, 44, 68, 99, 114
Soviet Union. *See* Russia
Spain: Basque guerrilla warfare, xviii; civil
war in, 82; designs on England, 55–56;
ETA, xxi; Inquisition, 21; in the
Philippines, xx, 35–41; Reconquesta,
24; Spanish-American War, 39–40;
xenophobia in, 24
Spanish insurrection: acquisition of goods
and resources, 29–30; anti-terrorists
(*chacales*), 31; battle of Saragossa,
25–26, 129; Catholic Church and,
21, 214; climate of insurgency, 24–26,
34; code of honor, 28; counterinsurgency in, 31, 229n; *cuerpos francos*, 31;
dismantling of guerrilla organizations,
32–33, 229n; formation of *juntas*,
xxii, 25, 32, 174; French brutality, xx,
24, 34; French invasion of Madrid,
30, 129; French strategy, 30; French
workers (*gavachos*), 22; growth of insurgency, 29–30, 34; guerrilla warfare,
xviii, xxii, xxiv, 11, 20–34, 129, 228n;
insurgency called "bandits," 16–17;

Mina as leader, 28–29, 30; Miquelet, 31; Mobile Musketeers of Navarre, 31; monarchy and, 21–24, 32; nationalism, 209; Navarre, 27–28; Pamplona, 29–30, 30, 32; sack of Cordova, 26; Second of May insurrection and Third of May retaliation, 24–25, 129; size of French forces, 30, 69; small numbers of insurgents and, 29–30, 33; small-scale arms industry, 29; terrain and, 27–28, 85; treatment of collaborators, 30, 210; treatment of prisoners, 32

Spooner, John, 40

Sri Lanka: guerrilla warfare in, xviii

Stalin, Josef, 82, 92, 88, 164, 183; deal with Churchill, 100–101, 238n

Stavrianos, L. S., 95

Stern, Avraham, xix

Stettinius, Edward, 159

Stevens, Sir John, 232n

Stiglitz, Joseph, 205

Stuart, William, 84–85

Sudan, 189; guerrilla warfare in, xviii

Sumter, Thomas, 11, 18

Swift, Jonathan, 57

Syria, 246n

Taber, Robert, x; xxiv

tache d'huile (oil spot), xxii–xxiii, 15, 153, 210

Taft, William Howard, 45, 51, 52

Taliban, xxiii, 201, 213–17, 258n

Tanganyika, 107

Taruc, Luis, 49

terrorism, xviii, xix, 31, 143; car bomb as weapon, 66; dominant government's response, xix–xx, 31; execution of opposition and, 13; insurgency becomes a coherent movement and, 79; organization of terrorist cells in pyramid, 140–41; *plastiquer*, 141–42; quasi-judicial stage, 13; small numbers and, xix; as tactic of, xiii, xix, xix–xxi; used against rival factions, 12, 14. *See also* IRA; *specific countries*

Thatcher, Margaret, 69

Thompson, Sir Robert, 118, 121, 153, 225n, 241n, 251n

Thomson, James C., 225n

Tierney, John, 209

Tito (Josip Broz), xx, xxvi, 30, 104; background and rise of, 80–83, 235n; British officers with, 83–85, 86, 100–101, 157–58, 234n, 235n; campaign against the Nazis, 130; Churchill's deal with Stalin and, 100–102, 238n, 239n; Communism and, 81–82, 98, 101; creation of "anti-state," xxii, xxiv, 29, 129; Greece and, 105; Partisans of, 78, 79, 95, 97, 139, 140; Stalin and, 88, 92

Tocqueville, Alexis de, 131, 243n

torture, 141, 143, 245n

Truman, Harry, 164, 167

Truman Doctrine, 105, 166

Turkey, 68, 100, 207

Turkmenistan, 253n

Ugly American, The (Lederer and Burdick), 50, 231n

United Nations, 68, 157, 159

United States. *See* America

U.S. Agency for International Development, xv

U.S. Department of Defense, xv

U.S. Department of State, xv; Office of Far East Affairs, 165; Policy Planning Council, xv, xvi, 144, 243n

U.S. Military Academy at West Point: text on warfare, 33, 41–42

Vafiadis, Markos, 104, 106

Vance, John, 121

Van Fleet, James A., 105

Vann, John Paul, 136, 251n

Viet Cong *See* Vietnam

Viet Minh *See* Vietnam

Vietnam, xxi, 156, 245n; America and, xv–xvii, 163–80; American advisors in, 157, 158, 251n; American "alternative strategy" in, 167–68; American backing of France, 164–67, 243n; American counterinsurgency, 153, 169–70, 175, 178, 200, 254n; American frigate fires on (1845), 150, 163; American policy, 50, 136, 153, 164–65, 166–67, 169; America's rejection of Ho, 166; attacking of fellow inhabitants and, xxi–xxii, 173–74, 210, 250n; Bao Dai rule, 148,

Vietnam (*cont.*)

159–60, 176–77; China and, 146, 148, 151, 160; Christian missionaries and Christianity in, 147, 149, 152, 156, 163; CIA (OSS) in, 157, 158–59, 163, 168, 169, 174, 179, 241n; Cochin, 150; Communist Party (Lao Dong), 151, 173, 175–76; concentration camps in, 155; "Confucian balance," 161; control of territory and, 13, 227n; corruption of government, xvi; cost, 122, 180; counterinsurgency plan, 118, 121, 136, 241n; creation of a climate of insurgency and Diem's regime, 171–72, 173; creation of alternative administration, 161, 248n; creation of permanently held rebel-controlled areas, 15; Diem, 168–71, 176–77; Dien Bien Phu, 136, 137, 157, 162, 167, 169, 187; domino theory, 99, 159, 164, 252n; DRV, 176; European arrival in, 146–47, 149; French brutality in, 24, 155; French counterinsurgency, 152–53; French destruction of indigenous culture, 153, 154; French forces in, 162; French occupation, 145, 150–62, 170; French repression, 153, 155; French strategy in, 30, 153; Geneva Accords and division of country, 167, 168, 176; Giai Phong Quan (southern Viet Minh), xxi–xxii, 173–74; Giap leadership, 157, 162, 169, 180, 229n, 246n; guerrilla warfare against the Americans, 178–80; guerrilla warfare against the French, 86, 151–54, 160–62; GVN, 178; history, 144–48; history of clandestine movements, 151; Ho Chi Minh and resistance to Japanese, 156–58; Ho Chi Minh leadership, xx, 148, 159; Ho Chi Minh Trail, 179; insurgency, issues focused on, 156; Japan's military victories, 154–55; land and its inhabitants, 145–46, 150–51, 172–73; Lansdale and Diem, 148, 169, 170; Lansdale and U. S. policy, 50; Malayan model, 118, 241n; military technology and terrain, 86; model for counterinsurgency in, 105; *montagnards*, 146; nationalism, xvi, xx, 161, 173; negotiating climate

and American withdrawal, 180; Nguyen Anh dynasty, 148; party structure, 8; pattern of insurgency in, 173–74; Phan Dinh Phung, 152, 153; Phase 2, xxi–xxii, 6; Phase 3, 174–75; Philippines model, 35; Phoenix Program, 179; political issue, xvi; Polk and, xv, xvi; as quagmire, 43; resistance movements, village oriented, 151; revolt of 1930, 156; Rostow and, xvi; size of insurgent movement, 96, 157; social organization and culture, 150–51; "stay behind" units, 168, 173; strategic hamlets, xxiii, 151, 172–73, 177–78, 179, 210, 251n; the *tache d'huile* (oil spot), 15–16; *tadien*, 246n; Tet offensive, 180; Viet Minh/Viet Cong, xvi, xix, xxi–xxii, xxii, xxiv, xxv, 16, 86, 136, 146, 156–62, 166, 168, 171, 175, 179–80, 199, 239n, 250n; war of national liberation begins, 154; World War II and, 156–58, 160, 164–65

Vincent, John Carter, 165

Viollette, Maurice, 132

Vitkevich, Yan, 185–86

von Steuben, Friedrich, 16, 226n

Washington, George, 12, 226n, 227n; Braddock's defeat and, 4, 227n; desire for a British-style army, xxv, 4–5, 15, 18–19; disdain for militias, 17, 18

Wehrle, Leroy, 225n

Wellington, Duke of, xvii–xviii, 29

White African (Leakey), 122

Willkie, Wendell, 157

Wind That Shakes the Barley, The (film), 232n

Wohlstetter, Albert, 225n

Wolff, Leon, 230n

Wolff, Robert, 88

Wolpert, Stanley, 186

Woodhouse, Montgomery, 97, 237n

Woolsey, James, 221

World War I, 85–86, 131–32, 155, 190

World War II: in Algeria, 134; American policy in Vietnam and, 164–65; Axis countries, 91; China in Vietnam, 146; Crete, 91; Egypt and, 92; fight for Greece, xix, 91–102; German invasion of Yugoslavia, 73–81; Japanese in the Philippines, 46–47; liberation of Europe, 100; military technology, 86; resistance movements, 97–98, 237n; Russian partisans, 85; Tehran Conference, 88, 238n; Yugoslav partisans, xix, 82–88

Yarmolinsky, Adam, 225n
Yu Chi Chan (Mao), x, 152
Yugoslavia, xxvi, 72–88, 234n; Allies helping Partisans, 83–86, 100–101, 157–58; alternative administration and Partisan industry, xxiv, 29, 86–88, 96, 129, 174; as Axis country, 91, 92; Cetniks, xix, 77, 82, 97, 101, 197, 225n, 234n; Churchill's deal with Stalin and, 100–102; climate of insurgency, 73–76, 80; collaboration in, 76–77, 210, 234n; Communism in, 81–82; creation of permanently

held rebel-controlled areas, 15, 86, 137; ethnic cleansing in, 75; foreign aid and, 86, 97; formation of state, 73; German occupation, 73–76, 92; German reprisals, xx, 76, 77, 83, 200; guerrilla warfare in, 76–78; kings of, 73, 88; Mihailovic in, xix, xx, 77, 82, 234n; *odbors* (people's councils), xxii, 86; overthrow of government, 73; Partisans, 17, 30, 78, 82–88, 95, 101, 130, 139, 199, 234n, 235n; physical and social geography, 72–73, 82; size of insurgent movement, 96; superiority of German force, 69; Tito in (*see* Tito); treatment of prisoners, 32; Ustase (the "rebels"), 73, 79–80, 234n

Z (film), 239n
Zahir, Mohammed, 193
Zakhariadis, Nikos, 95, 104
Zervas, Napoleon, 97, 140

ABOUT THE AUTHOR

WILLIAM R. POLK STUDIED AT HARVARD (BA 1951, PHD 1958) AND Oxford (BA 1955, MA 1959). At Harvard he helped establish the Middle Eastern Studies Center and taught history, government, and Arabic literature. In 1965 he was appointed a member of the Policy Planning Council by President Kennedy. In that capacity, he was in charge of planning American policy in most of the Islamic world and also served as director of the interdepartmental task force that helped end the Algerian war, negotiated a cease-fire between Egypt and Israel, and was a member of the Crisis Management Committee during the Cuban Missile Crisis.

In 1965, he resigned from government service to become Professor of History at the University of Chicago. There he also established the Center for Middle Eastern Studies and was a founding director of the Middle Eastern Studies Association. In 1967, he also became President of the Adlai Stevenson Institute of International Affairs.

Among his books on history, world affairs, and the Middle East are *The United States and the Arab World* (Harvard 1965, 1969, 1975, 1980, 1991); *Neighbors and Strangers: The Fundamentals of Foreign Affairs* (Chicago, 1997);*The Elusive Peace* (Croom Helm and St. Martins, 1979); *Polk's Folly* (Doubleday, 1999, 2000); *Understanding Iraq* (HarperCollins 2005, 2006); *The Birth of America* (HarperCollins, 2006); and, together with former Senator George

McGovern, *Out of Iraq: A Practical Plan for Withdrawal Now* (Simon & Schuster, 2006).

He has lectured at many universities and at the U.S. National War College, the Council on Foreign Relations, the Royal Institute of International Affairs, the Canadian Institute of International Affairs, the American Foreign Policy Association, and the Soviet Academy of Sciences, as well as many civic groups.

Mr. Polk is the senior director of the W. P. Carey Foundation.